Phoenix by Three

Written by

Ronald Ian Gray

Table of Content

Dedication

To my grandchildren Alex, Leigha, Taylor, Carson, and our newest Charlotte, they have been inspirations and make me proud, when they achieve incredible goals.

Also, many thanks to: Roxanne Ortega and Charlotte Scythes – advisors
Donald Boak – content review

All the Staff at the Corner Pub in Thornbury, Ontario, for their major support.

Characters

Aileen, James, and Amy Parker

Matthew, Louise, and Gwen Provost

Grayson, Amelia, Aubrey, and Abigail Loughty

Chief Collard and Deputy Black

Jocelyn and Austin Masters, - James' secretary and husband

Lorraine Sanderson – clinical nurse

Fire Chief Arnold Black

Principal Angela Rutherford

Mayor Albert Standish

Andrew – the security guard. Angel

Aleksander Rickoff – dangerous criminal

Stanley Reynard – the fugitive from the drug homicide

Thomas McCarrique – OPP commissioner

Jo-Anne Bradshaw – nurse hired to be with Amy.

Peter and Glenda Davidson – brewery manager and dress shop manager

Prologue

Three years have passed since Parkerton and its citizens dealt with a slate of issues from plant closures, drugs in the local collegiate, student deaths, and the murder of one of their students. The arrival of the Loughty family from the UK, their major involvement in the community's well-being, and the successful search for their wife and mother had involved many of the townspeople.

Parkerton was struggling to maintain the smaller town ambiance but was gradually losing out to external influences, some of which did not conform to their stereotype

The Loughty family had successfully integrated into Parkerton society and the twins have decided to stay in and work with the stable and dress shop to save money for college.

James Parker, Aileen's father, and Senior Manager of the facility, challenged his work teams to seek positive solutions to repel a corporate return. The teams attracted three new companies to town and are working to get permission for a fourth. These included a brewery, winery, and EV battery reclamation facility.

Aileen Parker and Gwen Provost used the graduation ceremony as their 'coming out event'. This was received in various ways by the audience; gays were a new phenomenon in Parkerton.

The girls opened their equestrian business thanks to Gwen's father and quickly became engaged, thus signaling things were changing in a big way for them.

Stanley Reynard, the convicted murderer, was nearing his first parole hearing.

The finder's fee for initial contraband from the first find at the Harrelson estate was delivered to Grayson. He and his family had decided to put to use most of the reward money for the construction of a local community park as a small token of their appreciation for being accepted into the town.

Things were moving in a positive direction except for some anti-gay sentiment surfacing as the girls became more prominent in the community and were proudly gay.

Chapter One

The entire town breathed a combined sigh of relief, as the past few years have been without significant incidents or issues and it was as if they were waiting for the next shoe to drop.

The global economy is in trouble with the outside world, impacting town businesses.

Still, overall Parkerton is doing well, with the three new businesses, the retirement community, and a proposal for a casino located between Parkerton and Ayelsbury.

The addition of more people to the community had strengthened and increased the overall tax base, allowing the town council to approve additional fire personnel, extra police officers, and state-of-the-art ambulance and paramedic services, but the influx of new people has brought larger issues.

Several shops were added to the town. A new bakery, neighbourhood diner and pub, liquor and beer stores, fine dining establishments, and finally a local newspaper.

An indication Parkerton was fast becoming a larger center was the addition of a Tim Hortons fast food restaurant, *a major Canadian staple*, and the potential addition of a smaller grocery store from one of the larger chains.

Other additions to town were technology firms bringing the latest Internet, Wi-Fi, and 5G technology, allowing shops and businesses to install ATMs. In support of this innovation, another bank in town actively addressed the ATM and debit protocols with customers and local businesses. It was a hard sell initially, but people were buying into the new procedures and moving into a new era.

During the three years since the post-pandemic graduation, 50% of the grad class of that year departed for greener pastures. Two Clan 8 gang members, the young people involved in the drug episodes, still lived with their parents. They failed to complete their community service, as required by authorities, meaning their drug possession charges remained on their records, leaving them difficult to employ locally. Graduations since that time have proceeded smoothly, with the number of students increasing exponentially as a direct result of the influx of new families. This has also added to an increase in volunteers for future events.

Townspeople were always excellent at volunteering as this was the advantage of a smaller population, but it seemed to be the same persons all the time who were charged with running the fairs, exhibitions, and major holiday celebrations.

All appeared well, but under the surface, tensions were building in several areas, and the anti-gay and LGBTQS+ sentiment was growing louder and bolder. The girls were waiting for the right/wrong time to display their Pride flag, as the gay issue had become significant for some of the more opinionated townspeople, especially for Aileen and Gwen, being the first and only gay couple. For them, this was no secret as their love was pure. They had no shame in expressing it out in the open having the support of their families at their backs. This had a great potential for fears of narrow-minded people to boil over. This clash of cultures and mores was not unique to Parkerton. The new people have added large divisions in perceptions, attitudes, and paradigms affecting their assimilation.

Seniors living in the retirement community, for the most part, were from the city. They brought their views of how small towns should be governed and their priorities, causing major friction. They were making demands based on "city lives" and being generally more financially secure than most locals, and some of their demands bordered on fiscal irresponsibility.

They demanded bus service from their neighbourhood to downtown and back, hourly. Also, the transportation should travel to Ayelsbury and Proctor twice per week.

They requested senior discounts in all the shops and to have people come, mow their lawns, maintain the gardens, and shovel the snow. This had not been negotiated as part of their condo association and now they needed to address these issues with their board.

People in Parkerton were dealing with different priorities like peace, family life, and a slower pace of growth, making the gap between what they see and what the "others" see a huge chasm.

Similar scenarios were playing out in other small towns in the region, driven by the price of houses in the city driving immigrants to the smaller, less expensive areas. A consequence of this migration was the rapid increase in the price of local housing, creating problems for the locals and their children to be able to afford homes and provide for their families.

As the town grappled with these growing tensions and conflicts, it was evident that a resolution was needed very soon to avoid further major issues. The new residents and long-time locals had to find a common ground and bridge their differences. The path to harmony and coexistence was uncertain and the people were looking to the mayor and the council to undertake this challenge.

Chapter Two

Aileen and Gwen are turning 24 and they have decided to hold their wedding in the summer, knowing it will be quite an unusual event, being the first openly gay couple wedding to be held in their small town. Their relationship has already been causing tongues to wag and discriminatory innuendoes to surface and have had a small impact on the success of their Equestrian business.

Aileen's mother, Amy, had been diagnosed with ALS, a terrible disease affecting all muscle functions, and Amy, normally determined and strong, took the news extremely hard, especially when the Doctor indicated the disease had been present for a while, already affecting her life expectancy, which can be as little as 2-5 years. The rapidity of the disease progression was putting a strain on the family and requiring James to hire people to look after Amy's dress shop and to remain with Amy during the day. Abigail Loughty, a close friend of the family was helping out at the shop, but they needed an experienced manager and would have to look elsewhere.

The dress shop had been needed in town as there were no places for women to access clothing unless they went to Proctor which was 160km away. It is a small, boutique-type facility handling excellent quality merchandise at mid-range prices and keeping abreast of the latest trends and standard fare.

James also needed to have a person come into the home daily to look after Amy; with Aileen spending so much time at the council, much to the chagrin of Gwen, available people resources were becoming in short supply.

The whole Aileen and Gwen relationship was wearing on James and he was worried about their future if/when Amy passed away.

The girls tried to keep things as normal as possible, refraining from hanging a 'PRIDE' banner at the stables and not raising the LGBTQS+ issues unless asked directly. This was still a small town with some old-world ideology and pushing the envelope would help nothing. The only items that would even hint at a gay scenario, in some peoples' eyes were the lavender and pale blue colours they selected for their business, making them identifiable with their horse blankets, riding gear, company stationery, and business cards—all in two colours with their insignia, the rising Phoenix as their identifier.

They were planning a wedding this year, but the strain of the business and the amount of time Aileen was spending with the mayor and town council put a strain on their relationship. They have arranged to take a time out and spend less time together for a while, giving themselves a chance to see how they feel.

They are having some serious issues, with Aileen spending a great deal of her time consulting with the town council on developmental affairs and neglecting her responsibilities as a life partner and business partner and now the daughter of a very sick mother, this is about to come to a head.

The love, togetherness, and intimacy they shared rapidly deteriorated and they seemed to be drifting apart rapidly and had very different views on their relationship.

A large equestrian show-jumping competition is taking place in Proctor next week, at the exhibition grounds as preparation for amassing points to take their place at the National level. Gwen was working hard to get everything arranged with no assistance from Aileen.

The preparation time required to get the horses ready, clean all the tack, have the horses shod, ready the trailer and get all the documents in place by herself has quickly become a very sore point. She had to solicit assistance from the Loughty twins, Abigail and Aubrey, who were already familiar with the stable operation, to look after the daily chores, exercise the animals, and ready the stable for new customers coming at 2 p.m. on Saturday.

Gwen sent Aileen a brusque text in all CAPS. Aileen has not been staying at their small flat above the stable, but rather staying with her mother in her old room and trying to help out but appearing to be deliberately avoiding Gwen.

The text indicated she was readying everything for the show and wondered if Aileen would ever turn up to assist, or if she was too busy with her voluntary council ambitions to remain a part of the business and especially their life together.

The show was taking place on a Thursday and Friday this coming week, which was very different from the normal weekend shows and Gwen was stressed out about the preparation and participation. Aileen responded to her text saying she was attending an all-day planning session on Thursday and couldn't make the trip but would support the three of them attending and wished them well.

Gwen was furious. She replied again, "**I NEED YOU TO BE THERE**. This show is just too much for one person to handle and we are also having new customers arriving Saturday to board their two young mares. If you can't make the show and are not going to be there on Saturday for me, I need you to re-evaluate your continued involvement in our life relationship and the business. The gossip around town is you are spending a great deal of time with the mayor's son. If this is the new Aileen, then we are through. Make up your damn mind and get your head out of your ass.

"Do you know what you are doing to us and the business? It's up to you, but I am prepared to cut and run if that is what you want. I will hire another person full-time or encourage Abigail or Aubrey to come on board. I am leaving here early Wednesday afternoon to travel, get everything in place and do a couple of practice rides, and I need your help."

"Enough is enough. I am at my wits end with you, your new attitude and your narcissistic ego trip doing a voluntary position so you can be seen by the whole town has gone too far." Was her hostile reply.

There was no response from Aileen.

Gwen had the twins come out early on Wednesday, spend a few hours helping her get everything ready, and then challenged them to tidy the stable. "Please let all the horses out for a run." She requested.

The twins, currently had lots of time on their hands as they hadn't decided on career paths yet. Abigail was still part-time at the dress shop, making some money; working with Gwen was a monetary bonus so they agreed.

Wednesday morning came, Gwen was up at 6 am and starting to get everything in order when she heard the hum of the Vespa cycle the twins had purchased to get them back and forth. She broached the question about them coming on board full-time for a while to see if they were interested and Abigail was thrilled, but Aubrey was still hoping to get accepted to a university and wanted to keep his options open.

They now had nine horses aside from Bluster, Aileen's horse, and Artemis Gwen's, so a lot of work was required to get them all out to the pasture, clean out the stalls, and polish all the tack for the show. Busy day all around.

Abigail asked the question and was sorry after, "Where the heck is Aileen and why is she not here to help?"

Gwen didn't want to get into it, so she replied brusquely, "I would rather not discuss that topic right now. Let's just get to work, OK?"

By the time Gwen was finished packing up the trailer, putting all the tack in the proper place to stay clean, and because money was still in short supply, she had decided to take the cot and small mattress and stay overnight in the middle stall, a very unusual situation.

This would end up being an almost fatal decision. Pushing the cot and small mattress into the front of the trailer, she needed to pack her go-bag and hit the road. It was not a long trip this time, but there was a lot of setting up to do.

They gathered the two horses, got them inside, and filled the bins with hay and water. All was ready to go, but she was apprehensive about tackling this on her own for the first time.

She bid farewell to the twins, texted her parents that she was leaving, and then sent a snotty text to Aileen.

I know you don't care, but I am loaded up and ready to leave. Enjoy your meeting and time with your new friend; maybe I will see you later.

Chapter Three

The drive to the exhibition grounds from Parkerton was uneventful and Gwen was becoming even more and more annoyed that she had no response from "Her Highness" while not anticipating all the work needing to be done at the showgrounds.

The older attendant at the gate directed her to their appointed stalls and asked her to try and get the trailer out of the way as quickly as possible, but when she explained she was alone, he said, "Sorry to hear that young lady, just take your time, I will pop over to see if you need any help. By the way, I want you to be especially careful. There have been some strange things happening around here lately. Several very expensive horses have been getting ill and, in two cases, need to be euthanized and some others are just ill and nobody knows why, so please keep your eyes open. OK."

"Thank you very much my dear and what is your name." Gwen wanted to say thanks.

"My name is Andrew. I will be out and about. If there is anything you require or have any questions, please just find me and we will get what you need."

She occupied three stalls, one for each of the horses and one in the middle for herself. Again, this will be proven to be a very poor decision. She stored all the tack, associated brushes, combs, hoof paint, and blankets. The horses were settled and then she strung her new Phoenix curtains in the lavender and blue. She was very proud.

She was struggling to unload the small bed and mattress when Andrew, the attendant, true to his word, showed up to help, "Thank goodness, " she said, "my partner was supposed to be here with me and it would have been easy, but she has other things to look after this time so here I am on my own."

He was very concerned and replied, "Well the barn crew will help you with some of the duties and I will make sure to keep an eye out for you if that is ok?" As I said before, "There have been some strange happenings lately?"

"Most gracious and appreciated, you are a true gentleman." She gave him a peck on the cheek.

Everything was in place and looked after, so it was time to saddle Artemis and head for the practice ring. Her horse was still skittish and misbehaving, but Gwen chalked it up to the confusion of the packing and the trip. After three circuits, Artemis was fine and they carried on for fifteen more minutes.

Artemis was bedded down and now it was time for Gwen to look after herself. She grabbed a change of clothes, toiletries, flip-flops, and a towel. Off she went to find the changing room but all the time while walking she had the impression someone was watching her so she began taking additional precautions when on her own.

Everything went well and back at her cubicle she arranged the bed, hung up all the clothes for tomorrow's ride, and set up the three motion sensors and recording devices that she placed in a secure location. These were primarily due to a series of thefts happening at other shows, so these would activate and record anyone or anything entering any of the stalls where the horses were placed or in the one, she would be using as home.

She was tired but hadn't eaten all day, so she decided to go to the exhibition's restaurant on the grounds to get some dinner.

The meal was excellent and there was some music, so she decided to just hang out for a little while and have a glass of wine and talk to the other riders and owners.

All was fine until a man dressed like a rider sat down beside her and began asking personal questions. She asked him to leave her alone as she needed some air and left, trying to still look composed. If this person continued to harass her tomorrow, she would report him to Andrew and let him handle it.

Once back at the stalls, she fed the horses and provided a special treat of oats, fresh water, and a stroke on the neck and said goodnight and best wishes tomorrow.

Gwen grabbed her phone and texted her parents to let them know she was fine and, against her better judgment, sent a note to Aileen and asked her to check with Abigail and Aubrey to see how things were going, and she would check in the morning for responses. Just before putting the lock on the stall door, she thought for a minute and then texted Aileen again.

"Not that you seem to care, but I am here with the mares and settled in and will be riding at 10 am. Hope you are happy with deserting me the way you have done, and I will expect you to be at the barn early Saturday morning to get the place ready for the latest customers. If you are not there – WE ARE THROUGH. "

With all that done and the locks and security cameras in place, she laid down for a much-needed night's sleep.

The alarm went off and she was up and headed for the washroom to get tidied and prepared for the day. She was excited to be riding again and had high expectations for Artemis, as they were the A-team, but was unsure about Bluster as she still seemed to be irritated and anxious.

Breakfast done. A warm wave to her favourite security man, and she was off to saddle up and do a quick warm-up. All was looking good and she felt positive so when she was called to the ring, she sat proudly.

The announcer came over to the speaker, *"Now representing the New Phoenix Equestrian facility from Parkerton are Gwen Provost and Artemis."*

The ride went very well, although a little slow. They finished with no faults as has been the case for her and Aileen for all the shows to date. She was tied for 2nd place, a great start to the show and the season.

The next ride was at 2 p.m. with Bluster and she was more concerned about this one.

Back at the stalls, while focused on polishing the tack, pressing her outfit, dressing up the hooves, and braiding tails, she didn't appear to see the man from the other night standing at the end of the row and watching her intently.

A short rest and it is nearing the time for her next ride. She saddled Bluster and headed for the warm-up ring, feeling a little less anxious but concerned as the two of them were not used to each other and they both seemed to be hesitant.

The ride went well, but Bluster refused at the final jump and she needed to back off and try again. The next attempt was successful, but the refusal dropped them down in the standings.

She had to show up for the ribbon ceremony for the first ride, so gathered up Artemis and headed back to the ring. As they all lined up, everyone was high in expectations.

When they next assembled for the division in which Gwen and Bluster rode, they were all called and lined up in front of the judges and they started down the list with Gwen wondering where they would end up.

The announcer called out, *"In third place and the White Ribbon is Gwen Provost and Bluster representing the New Phoenix Equestrian facility from Parkerton for a good ride."*

Gwen was almost in tears; she and Aileen had never placed less than 2nd at any show and never faulted, but there was always tomorrow and a different class.

Bluster was brushed, wiped down, and fed and both mares were down for the night.

She didn't want to text anyone but needed to tell her parents she was okay but didn't do as well as she thought they could, but a lot was going on in her life right now. She gave them a quick text and would reply in the morning.

A break. That is what she needed, so she went for a walk to spend some time watching the other classes with the huge jumps; they were awesome. Maybe she would get there one day, but Artemis was too small to attempt those jumps and she would need a larger animal.

Yea Right, she thought. Where was she going to get the 10K$ or more to pay for that addition to their stable? She assumed it was still **THEIR** business and started to cry.

Chapter Four

Gwen stayed and watched some of the senior riders and was amazed at the professionalism of both the riders and the horses; she knew the amount of work and major expenses involved in participating at that level.

She was envious.

It was getting later in the afternoon and knowing she and Artemis were scheduled to perform in the morning she decided to have an early dinner and get the gear ready for the morning. The barn boys would be ensuring that the horses were fed and watered and were safe so she could relax.

Nothing could be further from the truth.

As she was finishing dinner, she asked the waiter for a glass of wine but wasn't watching who delivered it to the table. After about 10 minutes, she began to feel very drowsy and unstable on her feet and found herself being led back to her stall by a tall male rider whom she didn't know but looked familiar.

That was the last she remembered until she awoke much later and was shocked to see the same man standing beside her with an ominous silence and staring down at her like a wolf looking over a carcass.

It was only then she realized her boots had been removed, her jodhpurs were down around her ankles, her jacket and blouse had been unbuttoned and the man was now trying to remove her panties.

The last thing she can recall is someone banging on the stall door and hollering, "What the hell is going on in there? Get out of there!" The person screamed the police were on the way.

It seemed like she had heard the voice before, and then all went dark.

Andrew, her favourite guard, had watched the man lead her away from the restaurant and was very concerned about her safety. He got cornered by a whining rider complaining about the service in his area and it took him longer to get to Gwen and her attacker.

The assaulter had opened the gate and tried to run, but Andrew grabbed him and tried to hold him for the police. The man was younger and stronger and was able to escape, but not before Andrew had a great description for the police.

Fifteen minutes later, the Ontario Provincial Police and an ambulance turned up at the stall, responding to the call from Andrew.

Staff Sgt. Monroe was asking the ERT people to get Gwen checked for rape and assault but first please cover her up. The ERT personnel administered a shot that brought Gwen back to a talkable state and asked if the police could ask her some questions regarding the incident.

"Yes," she said, "It was terrible."

Sgt. Monroe proceeded with the questions but was taking care to be delicate.

"What is your name and address?"

Gwen replied, "Gwen Provost, Phoenix Farms, Parkerton."

"Do you know where you are right now?" he continued.

"At my stalls within the exhibition grounds in Proctor." Her confusion was lifting.

"Do you know what happened at all?"

"Not really. I was finishing dinner and someone brought me a glass of wine. Then I started to get very drowsy and remembered a man taking my arm and leading me back to the stall. "Nothing after that?" he asked.

"Not really, except for opening my eyes and seeing the man standing over me and realizing that all my clothes were undone. Then I remember a lot of hollering from outside the stall and someone screaming What the hell is going on, and it went dark again."

"Can you describe the man for me?" asked the Sgt. Gwen stood up and said, "You might want to get the video from the cameras in the stalls; they are hooked to the motion sensor and sent to my phone."

"We need to take you to the local hospital to have you checked out and although we don't believe he had the time to rape you, we need to be sure for your sake. Ok with you?" Monroe asked.

The ERT people asked Gwen, "Can you please take a change of clothes with you and leave these at the hospital as evidence?"

"For sure." She answered, "Just give me a minute to collect them."

"Yes, now please can we go to the hospital." Sgt. Monroe was concerned for her welfare.

"Can someone here please watch the horses and my gear until we return?" Gwen was stammering.

"I will be here and take care of everything, young lady," Andrew piped up.

Sgt. Monroe questioned, "Can we please have your name and contact details?"

"Andrew Aingeal of Goodman Street in Proctor – AndrewA444@hotmail.com."

Monroe asked, "Gwen do you have someone we can call to come and be here for you as we know you were here all alone?" He was very concerned now as it is very unusual for riders to spend the night at the barns with the horses, especially by themselves.

Gwen provided Monroe with Aileen's number to call first and then her parents.

They proceeded to the hospital.

The ERT and the Dr. confirmed the fact that there was no actual rape, but all the bruises and a bruised rib indicated that there was a struggle and she is very lucky to be alive and still in good health.

Gwen was hesitant but eventually mustered the courage to call Aileen.

Aileen's heart raced as she answered the phone and heard Gwen's tearful voice on the other end. "How are things going? I am worried and we need to talk?" Aileen said, her voice quivering.

Gwen's sobs were audible as she replied, "I am at the hospital in Proctor I and was assaulted. Please come to the hospital and get me as I am scared and the horses are all alone and it's urgent." Aileen's hands trembled as she hung up and immediately called Gwen's parents to inform them of the situation. They too were on their way to the hospital, racing against time.

Aileen was a lot faster driver than the Provosts and made it there in just under an hour and a half as her thoughts and prayers for the love of her life were driving her to Gwen.

Aileen raced to the emergency room and her eyes met Gwen's, who was bruised and visibly shaken. Aileen and Gwen both broke down and began crying. "This was all my fault as I should have been with you and I promise we are always going to be together as we are better together. Please forgive me, darling."

Chapter Five

Aileen was collecting Gwen and getting instructions from the Dr. when Gwen's parents arrived, extremely upset and angry and barely looked at her.

Matthew, Gwen's father, confronted both of the girls and asked Gwen, "What the hell happened and why were you here on your own? Didn't you realize the danger you might have been in?"

"I'm fine Dad. This was not all Aileen's doing as I made the decision to be here on my own and didn't think it through. It was fortunate that Andrew, the guard, was watching out for me. He is my Angel. Aileen and I are working things out. We were not communicating well and I was angry and hurt, but we are going to be all right again."

Matthew began yelling at Aileen, "What are you doing here? Haven't you caused enough problems already with the nonsense at the town council? We have heard the gossip about you and Stephen, the mayor's son. Look at the damage your ego has done."

Gwen yelled, "Stop it, Dad. This is all over and done and I do not blame Aileen entirely, as I reacted in haste and allowed jealousy to take over my thinking. If you love someone, you have to trust them and I didn't at the time. We will work out all these issues when we get home, but for now, we need to collect the horses and all the gear and head home."

He calmed down and outlined the trip home for them. "Gwen can come with us, Aileen. You can take the truck and the trailer with the horses back to the stable."

Aileen inserted, "Mr. Provost please, you are forgetting we have three vehicles here that need to go back so we need another plan."

Matthew questioned Gwen, "Are you up to driving back home?"

"Yes." She replied. "Aileen and I need to go to the barn and get everything re-packed and then I will follow her home in the other car while she takes the truck and the mares."

"Is that okay?" she asked Aileen.

"For sure." Aileen was still crying. "We can take our time and get you home so you can rest and recuperate. I will look after everything at the stables to prepare them for tomorrow and tidy up **our** flat."

Gwen smiled when she heard Aileen refer to **their** flat and that she would look after everything. This was a great start to the next phase of their relationship.

Louise, Gwen's mother, spoke up, "Gwen, are you positive you want to return to the way things were as you are very angry and hurt?"

"Yes, Mother," she was crying now as well. "I love Aileen dearly and we can work all this out once we get back to our flat and have some time to talk. All relationships have hurdles and this was our first but probably not the last."

"Ok," said her father, "let's head out. Your mother and I will start back as we have some things at the office that need to be done and we will meet you when you return. Please be careful. Aileen, you look after our girl and your partner!"

Aileen and Gwen drove back to the exhibition grounds and Andrew was waiting at the gate waving so Gwen had Aileen stop and introduce them to each other.

Gwen said, "This is my guardian angel Andrew, and Andrew this is my life partner Aileen."

Andrew was smiling, "Good, there are the two of you now. Things can be done easier and you both can be safe. Will miss you and maybe see you next year."

As they approached the stalls, they could hear their horses calling to them. They recognized the girl's voices and were thrilled.

Aileen said, "I'll back the trailer over and you gather up all the light stuff you can. We should be out of here in an hour and on the way back to our home."

Once they arrived at their stables, everything was unpacked, saddles and tack stored, and mares in the back settled, fed, and watered.

Gwen had all the light gear unloaded and stored while Aileen moved all the horses to the pasture as they needed the exercise.

Once this was all accomplished, they headed up to the flat for their talk.

Aileen started, "I am so desperately sorry my darling. I let the council stuff blind me to what we have together. Do you still want to be with me?"

"Of course, I do, but what about this gossip about you and Stephen? Is that true?"

Gwen was becoming very red in the face and the jealousy was surfacing again.

"We had lunch together a couple of times and that was it."

He also cornered me in the upstairs meeting room and began putting his hands up my shirt and down my pants, telling me that he could cure me of being gay and that I would forget about you. I kneed him in the groin and slapped him so hard that his face was red the rest of the day, and he couldn't attend the rest of the session.

Aileen leaned across to Gwen and drew her in for a kiss. Gwen was hesitant at first and

then joined her in the most passionate kiss they had shared in weeks.

They spent the remainder of the early afternoon wrapped in each other's arms, crying, it was Gwen who broke the mood first and announced to Aileen, "Let's get married right away."

Aileen answered, "Hell of an idea. Where do we need to go to get someone to perform?

the ceremony?"

"Have no idea. We should ask the Priest and the Reverend first and see what they say

about performing the ceremony but not to expect open arms."

"When?" exclaimed Aileen.

"Monday, we ask, OK," answered Gwen.

"It's a deal," laughed Aileen. "This should set the tongues wagging in town again."

The rest of the weekend went very well. Abigail and Aubrey turned up early on Saturday morning and helped get the horses back into the stable, where they could be cleaned, brushed, and fed.

The new customers came at 2 p.m. as they had agreed and were very pleased with what they saw and heard about the Phoenix operation and philosophy, so dropped off their three animals.

The horses they left for the girls were slightly different grouping than they were used to handling as one was a young stallion, one a mare, and also a Welsh pony.

They explained that these animals were taken from a farm whose owners were arrested for mistreating animals and because of that, they needed special care and attention.

The stallion was a mixed breed of quarter horse and Arab and was 5 years old, but the pony was much older and could be used safely for the children. The owners would bring the vet reports as soon as they were available.

This added another dimension to their operation as all the animals to this point had been in good health and other than routine checks needed no special attention. They would need to develop a plan and need to check with their vet to get her opinion as to the health of the newest arrivals.

Chapter Six

All seemed quiet in Parkerton for a change, with the new Phoenix resource businesses doing well.

The bi-weekly newspaper was out and reporting on the activities in the area, this

was new as there had never been a paper in the area before and the locals depended on news from the outside world via city newspapers or the local gossip mill.

It was early on the next Wednesday that James, Aileen's father and CEO of the new company, received an urgent call from the acting CEO of the long-departed company ErgoPoly, who had dumped them years ago. At the time they had decided to pull out of Parkerton and made a deal with James and his team to allow them the opportunity to take over and operate the business, as well as address the personnel and housing issues.

The new CEO was insistent on needing an immediate meeting with James to revisit their

signed and sealed agreement. It seemed the success demonstrated by the new company and their ability to attract new business and successfully address the overwhelming housing issues had come to their attention and they wanted to revisit their decision and see about getting back into Parkerton.

James was quick to respond and said, "There is no way we are prepared to go back and redo this agreement. Our teams have worked hard to bring all this together and the fact that you were prepared to desert us and our town will always remain a sore point so goodbye."

While all this was happening in the office, the teams in the field had been able to secure.

Agreements with the local farmers and landowners to have them plant the malt and barley required for the brewery and also provide leased land for the planting of grapevines and the growing of special hops needed for their unique brew.

James was very distracted as he was dealing with Amy and her ALS, the strained relationship between Aileen and Gwen as well the frustrating intrusion into the business from the previous owners. He had Amy's long-term health conditions to consider and decided to hire someone to manage the dress shop.

He was becoming very depressed and felt alone in all these issues and the people calling to review the agreement made it worse. He couldn't allow himself to get ill or show Amy he was depressed as she was already showing the stress of the situation.

He needed to ask Aileen for help. This was not going to be good for her.

James called Aileen and Gwen and asked them to come to a meeting with him to discuss his current situation and ask for suggestions.

They promised to drop over and see him, but they had some stuff to deal with first, as they wanted to check with the Chief and see if anything new had turned up regarding the assailant. They had no idea about the news that was about to be delivered to them.

The girls have their own major issues when dealing with everything that happened, sorting out their issues, and deciding to get married. Aileen called the local priest and reverend, requesting their assistance with the ceremony, but neither of them would agree to take on this responsibility as they disagreed with their LGBTQ lifestyle and refused to perform the ceremony.

This was hugely disappointing to the girls as they and their families had been attending services in the churches all their lives and knew the Reverend and Priest personally. James was not going to be a happy camper with their decision.

They directed them to go to the town hall in Ayelsbury, register their request, and have a Justice of the Peace perform it for them. They will be calling Ayelsbury in the morning and hoping for the best.

While all this was happening, Chief Collard of Parkerton police received a call from the OPP in Proctor with some news and an update. Sgt. Monroe, leading this investigation said, "Chief, I have some good news, bad news, and weird news and would appreciate it if you would get hold of Gwen Provost for us. You can tell her that we had a suspect in custody regarding her assault, but the case has fallen apart.

The video from the stall camera did not show a clear picture of the man's face but did give us a great overall description.

We picked up a man in Proctor who has been suspected of other assaults, but he alleged that Gwen invited him to her location and that the potential for sex was her idea. We know this to be untrue as she was indeed drugged, but before we could get him to court, he disappeared.

The hospital still has her clothes and the samples Gwen gave them, and we have done a DNA analysis. The DNA is not in any of our databases and we have nothing to compare it with, so that is also a bit of a dead end.

I know she will be disappointed and angry, but we are out of leads for the time being. Sorry.

Now the strange news. We have tried to contact the security guard, Andrew Aingeal of Goodman Street in Proctor – AndrewA444@hotmail.com, and the name and e-mail address don't exist.

The security company responsible for the exhibition grounds told us they didn't employ an

Andrew is working for the event and they have never heard of him. They checked records back to 1920, the first year of the horse shows in Proctor, and found an Andrew, but the addresses were all phony there as well.

So now we have the assailant as well as the primary witness to the drugging and assault, both disappearing at the same time.

I don't know what to tell you other than we are still looking for both of them but the name,

address, and email for Andrew were all false. It gets stranger yet. The perpetrator has been identified as Alexander Rickoff as the prime suspect in the poisoning of the horses as well, and his address is fake. He is an average-sized man with a short beard and a mustache and is well-built, having black eagle tattoos on the back of both hands. His use of English was suspect and he had dark black eyes and a menacing posture when confronted.

The poisoning was apparently related to the female horse owners complaining to authorities about a strange man dressed as a rider lurking about and harassing riders. Shortly thereafter, the horses of the complainants were poisoned and notes were sent to their owners to back off, or they would be next. We are assuming that Rickoff was responsible for all these nefarious activities, displaying a lack of conscience.

Can you please relay this to Gwen and offer our apologies but tell her we believe her and are not giving up on the investigation?"

Chief Collard called Gwen and asked if she and Aileen could come to the station at 2 pm

that afternoon as he has some information to share with them.

"Sure." Gwen replied, "We will be there and hope the news is good."

2 p.m. and the girls arrived and were met by the Chief, who escorted them to his office.

"You had better sit down," said the Chief hesitantly.

The Chief took his time and relayed the points shared with him by the OPP.

- They had a man in custody, but he swears you invited him to the stall for sex.
- They checked with the security people regarding the witness, and there is no such person as Andrew Aingeal, and his address is phony.
- The suspected assaulter has a phony address, and DNA doesn't exist.
- Both of the people involved have disappeared.
- The camera footage wasn't clear enough.

The OPP expressed their apologies but were handcuffed by the lack of physical evidence and the disappearance of the people involved.

Gwen and Aileen were speechless. Gwen was crying, "So now what do we do, Chief? Does this maniac get away scot-free? Do we need to be afraid for our safety?"

Chief stuttered and replied, "I will have people checking in with you all the time, but make sure you have your motion sensors and video in place and lock all the doors."

Aileen commented loudly, "We know this is not your fault Chief, but it doesn't seem right. We can't continue to live like this long term."

"NO, it doesn't, and you shouldn't have to. But I promise not to drop this and to stay in touch with the OPP every week and share with you any new details." He replied.

The girls left in tears but knew there was nothing to do but wait and hope for justice to prevail.

They talked on the way back to the stable and decided that a trip to Proctor was in store for them and they would do some investigations on their own. Maybe the people would be more open to speaking with them than the police, but in doing this, they needed Aubrey to hold down the fort while they left for the day and also needed to spend some time with Aileen's father.

Chapter Seven

They met James at his office and heard what he had to say and how he was feeling about his ALS diagnosis for Amy, his depression, and his inability to help. He was struggling badly.

Aileen was becoming very concerned for his health and well-being and, after giving it some consideration, had a few ideas to share.

"Let's go to Amelia, at the clinic and see what she has to say about the ALS and its prognosis. Maybe she could make some suggestions for home care and medications."

She also suggested that her father take time out from his responsibilities as CEO of the company, reorganize his teams, and enable them to accept more ownership and decision-making responsibility. Her dad had just been maintaining an arms-length distance from the day-to-day activities and delegated some decisions to the people he trusted.

Gwen added that he might want to look at their home and see what immediate changes could be made to make the time remaining for Amy as comfortable and safe as possible.

Grayson Loughty, his 2nd in command, was a design engineer who graduated in the UK so let's get him to review the logistics. We should also contact the provincial ALS society and solicit their advice as well.

James was amazed at the suggestions coming from the girls and had underestimated their maturity and insight. "Thank you both," he was tearing up, "I will look into all of these, but would appreciate your assistance with the ALS society contacts."

"No problem," answered Aileen. "We are off to Proctor tomorrow morning for some research and will stop in to see them. Please stay as positive as possible and we are in this together."

When the girls returned to their flat, they discussed the next steps in both problems and came up with an action plan. Gwen would research the address given by Andrew and see which company provided the security people for the exhibition, while Aileen would call the ALS people and get a time for an interview.

First, they needed to contact Aubrey to make sure he was available and complete their duties to the horses and ensure they were all well and out to pasture.

They were leaving early the next morning and were very quiet for the entire 1.5-hour drive. The first stop was at the address provided by the mythical Andrew to see what they could learn, and then off to the exhibition security office.

The address was hard to locate. An obscure road, off the beaten path with only four houses on its whole length. They knocked on the #44 door and it was answered shortly by a small, grey-haired senior who was not too happy about their appearance.

"What the hell do you two want?" she exclaimed.

Gwen replied, after being startled, "We are looking for Andrew Aingeal and wonder if you knew him?"

"NO." was her terse reply. "The police were here last week and we told them the same things. No Andrew here, never has been and we know nothing. Now get lost."

"But he gave this as his address when working at the exhibition grounds," Gwen countered.

"Don't care what the person gave. We have never heard of him. Now please leave before I call the police." She was furious.

The girls departed, very discouraged, sensing the woman knew more than she was admitting and this piqued their curiosity even further.

The next stop provided even less information as the security people refused to release any personal information to anyone but the police. They admitted the fact there had never been Andrew Aingeal working for them and the only Andrew they had on the books was back in 1924.

Stumped again and frustrated, they headed to the ALS Society and looked for insight and assistance.

The Society provided a wealth of information and the appropriate contact details for James to call and get much-needed advice and assistance. Their group is prepared to come to Parkerton, visit with James and Amy, and make recommendations, as well as provide financial assistance for necessary mobility items and access to the house and bathrooms.

"Thank goodness," Gwen announced, "at least some success in our journey.

When they arrived back at their car, they saw a piece of paper under the wiper blade. Aileen was annoyed, "A frigging parking ticket. Damn. Wait that can't be. This is a private spot for the ALS Society patrons."

She grabbed the paper and when she read it, she was shocked. "Look at this she called to Gwen."

Gwen read the note out loud.

"IF YOU DON'T STOP STICKING YOUR NOSES WHERE THEY DON'T BELONG YOU WILL BE VERY SORRY. NOW GO HOME."

The next stop was the OPP station on their way out of town, where they asked for Sgt. Monroe.

He came to the front desk and asked what they were doing in Proctor. They explained their mission and then showed him the threatening note. Expecting some little empathy, they were surprised when he reprimanded them for interfering in an active investigation and they needed to stay out of it and let the police handle it.

Gwen was furious, "Well you don't seem to be doing such a great job, do you? Losing both people who could testify and not being able to track them down. I was the one attacked and nobody seems to care about me."

Sgt Monroe was embarrassed and apologetic but reiterated the fact that they needed to go back home to their horses and leave the investigation to the professionals.

"Fine." Said Gwen, "This is not the last of our investigation as I want answers."

The girls left and headed home frustrated, humiliated, and angry.

They needed to stop and see James and give him the update from the ALS Society and maybe some of that would relieve a little stress and depression, although this has been, and will be, trying times.

Once they had completed their update to James, they stopped at the police station to give Chief Collard an update on the way they were treated in Proctor by the OPP Sgt. Monroe.

Chief Collard looked at the note and asked them, "What the hell were you doing in Proctor?"

They explained their story and what they were trying to accomplish and to their amazement, the Chief sided with the OPP. "Stop what you are doing right now and concentrate on your business, your father, and your future. Apparently, you have rattled the wrong bushes, and I don't want you girls to get hurt or worse."

Aileen was fuming, "What the hell is going on Chief? Everyone is holding back information and it is infuriating us. We are not children; we are tax-paying adults."

"I am not entirely certain what is happening, but you are obviously in over your heads and we need you to stop for now." The Chief was worried.

The girls said, "As if this isn't enough for us all to worry about, Stan Reynard, the convicted murder suspect, has been released on bail just the other day, and already two of the old Clan 8 members, the non-convicted in-crowd drug users from the collegiate, have been harassed while at university. Taylor Spencer-Gray and Leigha Roberts-Hossack were at different universities, but both had their cars vandalized. Someone had spray painted – **murderer** – on the sides and the whole university was gossiping about the incident.

"This is another major source of concern for us. Please just leave this with me and concern yourselves with your future."

Chapter Eight

The girls stopped off at Amy's dress shop to see how Abigail was making out and the place was very busy.

"How are doing Abby?" asked Aileen.

"It has been long days and I need some assistance. Can you get me some help and hurry?"

Aileen said she had an idea and would get back to her in the morning. "Keep the faith Abby. We have your back."

Aileen called her father and asked for help. "Daddy, can you petition the people in your organization to see if there might be a lady with retail shop experience who would be willing to assist Abby before she has a meltdown?"

"Will do Princess. Get back to you tomorrow."

Time for the girls to complete their chores that had been neglected and get the horses into their stalls, as there was a storm on the horizon and the horses would spook in the pasture.

When all was completed, it was time for dinner and a long talk.

When they were cooking, they kept looking at each other and knew they were thinking the same things.

"Do you believe the crap we heard today?" Gwen asked Aileen.

"It's insane. Everyone is hiding something and as the victim, you should be privy to the information about your case.We need to keep questioning and digging and not let this go away," Aileen was insistent.

"Are you and I okay? That is the most important issue?" queried Aileen.

"Yes, my darling, we are back together again and can handle almost anything that can be thrown at us." Answered Gwen. "Oh, by the way, let's remember to call the town hall in Ayelsbury tomorrow and get a date for the wedding ceremony."

"I will take charge of that." Aileen insisted.

Time for a snuggle and a good night's sleep.

It was 2 a.m., when they heard a commotion arising from the stable and bad memories of the last time this happened, with Stan hiding down there, his gun in hand so neither of them moved.

"Just the storm causing the horses to be restless." Said Gwen with some hesitation in her voice.

"Should I go, see?" Asked Aileen

"**NO**. Please don't leave me alone tonight as I need you close and be my love partner again." Gwen pleaded.

"Fine, I will wait until morning. Let's get some rest, so come over here my love, and hold me close. I have missed you terribly." Aileen was moving closer to Gwen and grabbed her into a protective embrace. I love you." She said and kissed her passionately.

Aileen awoke early and was cleaning up the flat when she heard a commotion again, and this time, she was sure it was the horses. What had them so spooked as the storm had passed during the night?

Making her way down the stairway she screamed at the top of her lungs.

"Gwen, get the hell down here and bring your phone."

"Coming," said Gwen. She didn't get the words out of her mouth when she saw what had alarmed Aileen and she panicked.

"What is that?" she asked.

"There's a naked person tied to the stall, with a black hood over their head and a note pinned to the hood." Aileen was frantic.

"Who could it be?" Gwen almost fainted as memories of the assault on her came flooding back.

"Take the hood off. "Gwen pleaded to Aileen.

"Why me?" asked Aileen.

"Because you are the strongest personality of the two of us, please hurry as we need to know who it is and if they are alive or dead." Gwen was struggling to maintain her balance.

"OK." Aileen relented.

Aileen removed the hood and there was a naked man beneath, with a gag in his mouth. His eyes were open and he was breathing, there was the word RAPIST written in blood across his forehead.

"Read the note." Pleaded Gwen.

Aileen opened the envelope and peered at the note but seemed to be confused. The note read:

Here is your attacker and the person responsible for the horse poisonings.

I believe he is also a human trafficker and a drug smuggler. They have been using horse trailers to do their dirty work. Do with him what you wish, but there is a special pill in the envelope, and if you make him swallow, it all will be over.

Signed Angel.

Gwen was startled and exclaimed. "I knew Andrew existed and it was not my imagination."

"Now what?" asked Aileen.

"Call 911 and get the chief here ASAP." Yelled Gwen.

Aileen called and immediately had the Chief on the phone. "We need you out here right now. There is a naked man tied to one of our stalls, with a hood over his head with a note for us. **We need Help."**

The Chief couldn't believe what he had just heard but considering what had been happening lately with the girls, he grabbed an officer and scrambled into their car, heading for the stable. As they were leaving, they called the clinic to get the ambulance and requested Amelia to follow them.

The last time the girls called from the barn, Stan, armed with a rifle, was holding the girl's hostage, and now this, what the hell is happening in our small town. Either the stable or these girls must be cursed, the Chief was thinking.

The Chief and Officer arrived in 10 minutes and he was amazed at what they encountered.

"Who is it? He enquired.

"We don't know. Read the note," answered Gwen, "and maybe that will supply some answers."

Once he read the note, he had several more questions for the girls.

"How long has he been here like this? Do you have any idea as to who it might be? Did you touch anything? Where is this pill that is described?"

Aileen tried to answer the questions and spoke.

"We have no idea how long he has been there, but there was a commotion around 2 am and we just figured it was the horses, so we went back to sleep.

No. We have no idea who he is except that the note says it was the man who attacked Gwen on the exhibition grounds.

No. We only removed the hood to see if the person was alive and the pill is still in the envelope, we have no idea what it is either."

Chief proceeded to untie the man and remove the gag while the officer had his firearm ready, just in case.

"Stand up and turn around. You are under arrest." The Chief announced and put the man in handcuffs.

"What is your name and what are you doing here? How did you end up like this? Do you know these young ladies?"

"I will answer none of your questions without a lawyer, as that is your way, but my name is Aleksander."

Both of the girls jumped back, gasped, and together said out loud.

"That is one of the men the OPP are searching for, and the other is the composer of the note. You need to call Sgt Monroe in Proctor right away." Aileen was insistent.

Just then Amelia and her ER assistant, along with the ambulance, turned up at the stable and attempted to do a preliminary assessment of the prisoner. He was angry and struggling violently, wanting no part in their intrusive behaviour, and uttered many threats against everyone. "I will kill all of you," Aleksander was swearing.

Amelia just looked him over briefly and aside from some bruises and rope burns, he looked well enough to travel and stay in custody at the station until the arrival of the OPP and allow them to take control.

The OPP was contacted by phone and they were adamant that the prisoner be guarded 24 hrs/day until they arrive and for the officers to be armed and careful as this person is considered to be a very dangerous International criminal.

The Chief informed the OPP that this was a small-town station having only a single cell, but he would endeavour to take all precautions and have an armed officer on 24hr duty.

They headed back to the station to await the OPP appearance in the morning.

Chapter Nine

Gwen and Aileen insisted they be present when Sgt. Monroe arrived at the station. They had a few comments waiting for him, and most of them were not very ladylike.

The Chief said, "Are you sure you want to be there?"

"Yes," Gwen replied. "They lost him once and if something goes wrong, we are taking our story to the National newspapers."

"The OPP is going to be here tomorrow at 10 am to pick up the prisoner, so will you be able to get there?" Chief replied.

"Nothing can keep us away," Gwen answered abruptly.

It was 0945 am and the girls were already on their way to the station when they received a call from the Chief and Gwen put it on speaker.

"I have bad news. "Chief Collard was almost apologizing as he spoke. "You don't have to hurry to get here as the prisoner escaped and put my on-guard officer into the hospital in serious condition."

"If this is a joke Chief, we do not find this amusing. "Screamed Gwen.

"Not a joke, ladies. He faked having a heart attack and when the Officer went to see what the problem was and to render assistance, he attacked him, beat him over the head, tied him up, stole his weapon, and escaped."

"I will have to explain this to the OPP when they arrive, but like yourselves, I need some answers. Who the hell is this person and just what is going on? He is a violent offender with military-type skills and training and has injured my Officer and I want/need some answers. "

"**SO, DO WE**." Hollered the girls, "We are still coming to the station and will be there shortly."

The OPP, Sgt. Monroe arrived 10 minutes past 10 am and began screaming at the chief, calling him incompetent and useless for allowing Aleksander to escape under their watch.

"What the hell kind of Mickey Mouse operation are you running here, allowing a prisoner to escape while being under armed guard?" Monroe was beside himself with anger and sarcasm.

The Chief responded calmly, "Just a minute Sgt. Didn't this man escape from OPP surveillance and custody, not so professional considering the massive number of officers you have at your disposal?" Chief was not taking crap from the OPP lightly. "We have a small station and one 2-year veteran; at no time did we expect to be dealing with a highly trained soldier. So just keep your sarcasm to yourself. If you hadn't lost him in the first place, this would never have happened. I have a badly wounded officer in the hospital and do not appreciate your attitude."

Sgt. Monroe apologized, but the embarrassment of the situation was obvious as he needed to report all this to his superiors and there would be major repercussions.

Aileen interjected, "We want you to tell us exactly what is going on here and why is this man so important and so damn dangerous?"

Sgt. Monroe answered. "Sorry people I can't do that, as it is classified and you people don't need to know."

Gwen jumped in, "Don't need to know. Is that so? Then maybe the Telegram newspaper in Proctor or the National news would be interested in the story and to hear about all your failures."

"You wouldn't do that." He responded.

Gwen answered and was fuming, "You're damn right we would. I was attacked. We found a naked man tied to our stall in the barn, delivered by a man calling himself the Angel, who was looking out for me at the horse show and saved me from getting raped. So, I believe we deserve some answers.

You told the Chief and us nothing about how dangerous this man is and why he was wanted. I'll just bet the papers would love this story. Now either you tell us what the hell is going on OR we go to the press."

Sgt. Monroe opened up. "OK alright. I will share with you what I can, but this information is strictly confidential and is to be shared with nobody and I require your solemn oath to make that crystal clear.

"He is Aleksander Rickoff a small cog in a larger wheel of corrupt activities ranging from trafficking in drugs, smuggling large quantities of drugs using horse trailers as carriers, to human trafficking. We were trying to get him to be our CI 'confidential informant', but his people got to him first and threatened to kill him. It scared him and he was running from them and us. "

"Now we have a terrified antagonist on the loose, with a police weapon and apparently no vehicle.

The fact that he escaped and is on foot makes him an even more dangerous suspect in search of transportation and a place to hide.

The fact he knows about your stable and its isolation from town worries me, and we need to position undercover officers and car outside your establishment until we capture him or at best know he has left the area."

"Well, that is just freaking great, isn't it? What about our safety and the safety of our animals and customers?" Gwen questioned. "Do we have to alert all the people having horses with us and have them stay away until you tell us otherwise?"

Monroe answered, "Yes. That would be a good idea, but we need to come up with something that will not scare them, just have them stay away for a bit."

"What do you suggest?" Aileen was getting furious.

"How about a Covid alert cover story for a week or so?" he answered.

"That might be a short-term solution," insisted Aileen, "but the town gossip mill is going to have this information before morning, as gossip in this town spreads faster than the internet."

Gwen volunteered. "This arrangement was dubious at best and how do we know he is not out at the stable right now waiting for us and wanting to take our car or truck?"

"We will send officers back with you to do a complete search and also install more video and motion sensors around the stable."

Sgt. Monroe was suggesting something to try and alleviate their significant and real concerns, but it wasn't helping a great deal.

Aileen and Gwen were escorted to the stable by the OPP and two appointed officers, who appeared less than thrilled with their babysitting role. They commenced a complete search of the entire facility upstairs and throughout the barns.

They found nothing out of the ordinary and proceeded to install a couple of CCTV cameras and two more motion sensors and connected them to the girls' phones and the OPP office.

"We are just heading back to the station for a personnel change and there will be one officer all the time for the next week." The Corporal outlined the process for their return.

What they didn't see was the envelope stuck to the tack room wall addressed to *my favourite equestriennes*. Gwen opened the envelope carefully and showed Aileen the contents.

The note read:

I have been watching everything that has been going on and when the police leave, I will knock on the barn door 5 times as a signal. It is okay to let me in.

Signed – Your Angel, Andrew.

Gwen breathed a sigh of relief and spoke. "See, I knew he was no ghost and it will be grand to be able to say thank you to my Angel."

Aileen questioned her, "Are you sure this is the same man who came to your rescue?"

"We will find out when he knocks, right." Gwen was smiling.

A short time later, the knock came on the door and it was just as Andrew said. 5 times slowly. Gwen rushed to the door and flung it open and then stood there in amazement. It was him. It was Andrew. She reached out and gave him a monster hug and a kiss on both cheeks.

Andrew was not like they had imagined him to be, or even Gwen remembered. He was a man in his late 50s with deep blue eyes, a head of dark black hair, and a well-trimmed mustache. He had a military air and was extremely well built, looking more like one of the characters from a spy film than a security guard.

"Look, Aileen, it's Andrew my saving Angel. I knew you were real and not a figment of my imagination." She was blushing.

"What are you doing here sneaking around in the bushes? The police were just here and they have lost Rickoff again, but I am betting you already knew that, don't you?" she was rambling.

Andrew explained, "I have been watching the whole episode from the time they had him in custody in Proctor until he escaped. I knew he would be dangerous and that you could identify him, so he was probably heading your way. I followed his trail and found him lurking outside the barn and when he wasn't looking, I knocked him out, stripped him, gagged him, put a hood on his head, and tied him to the stall. The rapist's note was an afterthought."

"The OPP told us their version of what is happening and is yours the same, WHO are you?" queried Gwen.

"My name is indeed Andrew and I work for an agency responsible for identifying, locating, and eliminating threats to Canadian National Security when groups or individuals are contravening our laws about Immigration, drugs, and/or trafficking of human beings. Anything more than that I am not at liberty to share, sorry."

"So, are you the Canadian version of James Bond then?" was the question from Aileen.

Andrew laughing, replied, "Not quite. We are just people who stay in the shadows and make sure everyone is safe from these crazies. I will be keeping an eye on the two of you for a little while until I am satisfied you are safe. You won't know I am around but I will leave you envelopes from time to time. If you see that lunatic around, here is my special number, only for emergencies OK."

"Now you two go about your business and get on with the marriage plans. Goodbye for now."

"How the hell did he know we were planning on getting married soon?" Gwen was amazed.

"I don't know or care, but I definitely feel safer and more secure so let's take his advice and get in touch with the Justice of Peace in Ayelsbury so we can officially be life partners or Mrs. Parker and Mrs. Provost. We can kill two birds with one stone as we can go to Ayelsbury and talk to the JP and then continue to Proctor to the ALS people and pick up the accessories from them for my mother. What do you think?"

"Super. Let's go tomorrow afternoon, but we need to call the JP and the ALS people to make sure they can accommodate us." Gwen was giggling.

The next morning, they called the JP to see if she was available to talk in the afternoon, but the ALS people informed them they could get all the appliances we needed by going to the hospital in Ayelsbury and ordering through them, it would be closer and more efficient.

"Amelia, Aubrey's mother is the hospital administrator here in Parkerton, so let's speak to her first." Aileen was thinking ahead.

"Great idea, the less driving, the better." Said Gwen." We need to go to a jewelry store and order our two gold bands once we can get a date for the ceremony." Now Gwen was thinking ahead as well.

"Are we going to tell anyone?" Asked Gwen.

"**NO**." Aileen was adamant. "Everyone has enough in their lives happening right now so we can keep it for a surprise later."

"Good idea," Gwen replied. "I love you a thousandfold, you do know that don't you?"

"Of course, I do silly and I want to spend our lives together. Now let's eat something." Replied Aileen.

Two weeks had passed and things seemed to be returning to normal, or as normal as possible considering all that had transpired.

The girls had called the JP in Ayelsbury and scheduled a time next week for their wedding. They needed to get to the small jewelry store beside the dress shop, which would make it visible for all to see, but to hell with them. We are ready and can wear our special outfits saved from the graduation in lavender and pale blue. They planned with Aubrey to come to the stable and look after the place for the day while they were away, so the horses were in good hands.

They contacted Amelia at the clinic and she is ordering the appliances for Aileen's mother and they will be available at the clinic in a week. Aileen's mother seemed to be going downhill faster than the Dr. had indicated and her father was getting frantic about the accessibility at their home. He and Grayson were devising a plan. Things were going well for now, but a brewing storm of uncertainty and desperation lurked just out of sight.

Chapter Ten

There had been many suspicious incidents in the area lately and everyone was on edge again. Stan Reynard, the parolee, had been spotted at the universities and was suspected of spray-painting murderers on the cars of the 2 Clan 8 members attending different schools.

The Clan 8 had dispersed or had been accounted for as 2 were deceased from drug overdoses, 2 were living at home and still on parole, 2 were at university and 2 had just left town together and hadn't been seen or heard from since.

Stan was still demanding retribution for the death of his sister; he never believed his sister to be the founding member of Clan 8. He still was convinced they had conspired and were responsible for her death.

This had been proven in court to be false, but he was psychologically unbalanced and would not give up his harassment. The police have been informed of his return, not returning to Parkerton was a condition of his parole, and he was long past caring.

There had been a couple of strange fires in the downtown area and Chief Collard initially believed it was kids, but now they are focused on someone deliberately setting them, an arsonist, but to what end? Was it meant to be a diversion?

The absolute worst thing happened as the other day, while cleaning out the stalls and taking the horses to the pasture, the girls were sure they saw someone lurking suspiciously around the barn but magically disappeared when they saw him. This was not good.

Aileen asked Gwen, "Do you think it is time to call Andrew again and have him come to check things out?"

"No, not yet sweetie," Gwen replied. "We don't want to look like panicky women and cause a stir until we are certain. Remember the story about the boy who called wolf and nobody believed him? We are adults and can look after ourselves until we need Andrew, and hope it never comes to that."

They had been so preoccupied with everything that had been happening and they had been neglecting their business and the horses' need for attention, all of them. They spent the rest of the day cleaning stalls, giving them fresh straw for bedding, and replenishing their hay, water, and oats.

When they had completed all of that, they were going to turn them out to pasture for the day but decided to saddle up Artemis and Bluster and go for a short ride. Many of their customers were coming out this weekend to attend to their animals and go for rides, so they needed to ensure the trail was open and nothing was going to cause problems.

While they were out riding, down by their pond, Gwen suddenly pulled up and yelled to Aileen, "Did you see that?" she screamed.

"See what?" answered Aileen.

"There was someone back there in the trees watching us and when I turned around, they ran away." She was panicking at this point.

"You sure?" Asked Aileen. "Let's go back to where you saw them and see what we can find. Ok"

Once they returned to the spot, Gwen was sure she saw an intruder. They dismounted and decided to look around for any signs of a trespasser. They found an area that had been trampled down, branches bent, and several strange cigarette butts on the ground. They took photos of the area and picked up the butts using a piece of cloth Gwen had in her pocket. Not wanting to destroy any evidence.

"When we get back, I will call Chief Collard and show him what we found. OK." Aileen was also getting concerned.

They raced back to the stable, put the horses away, stored the tack, and went up to their apartment.

The place had been trashed, things were strewn all over, drawers emptied, and clothes were thrown everywhere.

"OK," said Aileen. "Leave everything the way it is and I will call the Chief to come right out and see what we found and the bloody mess." Was this Aleksander, Stan, or someone else? This was just getting scary now.

She called and told the Chief what they found and the mess in the apartment. He agreed to come out right away.

"YES." Gwen was shaking. "We can't allow these lunatics back into our lives, not when things are going so great."

The Chief arrived 20 minutes later and had the officer with him, who had been attacked by Rickoff at the station and was still recovering from the beating and the humiliation. He was very young and inexperienced and the girls knew him from their graduation class; he was David Clarendon.

They scoured the area where the girls found the evidence and then returned to the stable to look at the mess in the apartment.

"This was no random attack or intrusion." Said the Chief. "You are being stalked and the cigarette butts you found have Egyptian Cleopatra markings, which would lead me to believe we can rule out Stan as the perpetrator. That only leaves some unknown person or Aleksander Rickoff," he continued.

"I believed that Rickoff had disappeared and nobody knew where he was right now. I don't have the manpower to place someone here full-time time so I am suggesting you both go and stay with your parents until we sort this out."

"**No way.**" Said Gwen. "**We** are not children and have many customers. We need to make sure our CCTV cameras and motion sensors are working and phones are always charged. We can't just run away and leave the horses and customers to their own devices. We will make a concerted effort to be more careful and will not attend any shows until we get a resolution. Are you in agreement Aileen?"

"Yes, my angel, I agree with you and I am not willing to just cut and run, hiding like a pair of frightened rabbits."

The Chief was not happy with their decision, but he didn't have the people to position someone here permanently but would have Officer Clarendon patrol every 4 hours.

"You make sure you call me at the slightest hint of trouble. Do you understand?" The Chief was furious at them but terribly concerned.

Once the Chief left and was out of hearing range, Aileen said. **"NOW,** we call Andrew, OK?"

"Please do." Gwen was shaking. "He might be the only one who can help us now."

Aileen called and the phone went to voicemail and said, "You have reached 555-9999. Leave a message and I will call you back ASAP."

Aileen cried the message, "It's Gwen and I. We need your help. Andrew, please come, as Rickoff is stalking us again and has ransacked our apartment. Help us please."

The girls didn't expect to hear back for a while as there was no knowing where Andrew and his undercover role had taken him so they went back downstairs and started to oil and clean the saddles and the tack to take their minds off things for a while.

While waiting, they realized the horses were due to come back inside and get cleaned up so they went to the pasture and called Artemis.

The rest of the herd always followed her and pretty soon all the horses were brushed, watered, and given a treat of horse apples collected from the local orchards.

Suddenly there was a loud knock on the barn door. A distinct 5 loud knocks.

"That sounds like Andrew's signal." Hollered Gwen. She ran to open the door and there was Andrew, as large as life, with a huge smile on his face.

"The cavalry has arrived." Andrew's smile diminished. "Has he hurt my girls?" He asked.

"No," they replied in unison. "But he is scaring us and he just keeps hanging around and it is threatening." Aileen was shaking.

Gwen went to Andrew and gave him a huge hug and a kiss on both cheeks. "Boy are we happy to see you, Our Guardian Angel."

"Wait." She injected. "Did I feel a gun when I hugged you?"

"Yes," Andrew replied. "It is not unusual for idiots like him to be carrying a weapon so I armed myself. Now what about those cigarettes you found?"

"The Chief took most of them, but we kept two, as we thought they might mean something to you," Aileen answered and went for the butts.

"They are his all right. Those smelly Egyptian ones are not available in Canada, so it appears he must be around somewhere. I will begin my investigation and report back shortly, so please ease your minds as I am on the case but be careful regardless."

The girls were supposed to go to the JP that afternoon and were conflicted as to what to do when Andrew said to them, "Go, get married and I will be a witness if you like. Life has to carry on and you two are so made for each other. **SO, GO**."

They decided to go and stop at the jewelry store to pick up their rings. The owner of the store came to them and asked, "Is this the big day, and have you told anyone?"

Aileen said, "No, there is just too much happening but we didn't want to wait any longer, so we are going."

"Good for you." The owner smiled and said, "I will keep your secret, trust me."

"Thanks." Said the girls, and away they went.

The Justice of the Peace, a young woman in her late 20s who had only just received her designation was excited to be able to do the ceremony for them. She was very caring and thoughtful and went through the whole ceremony with a smile on her face. The exchanging of rings and their kiss started all three of them crying. They signed the register and all the papers the JP gave them, and then her last words, "This is only my second marriage and my first gay marriage and I am so proud of the two of you. Today is a milestone for me, the two of you, and the town of Parkerton so make us all proud and become the best and happiest married gay couple in the province. Go with my Love."

They went back to their car, holding each other tight. Aileen said to Gwen, "Well Mrs. Provost, how do you feel? "

Gwen responded with a passionate kiss and replied, "I'm so happy. What about you Mrs. Parker?"

Aileen smiled and said, "I am ecstatic and love you dearly. I have a question though. Did you get the same feeling from the JP that I did? That was she felt more in tune with the two of us and our struggle than most people."

"Yes." Replied Gwen, "I do believe she is a member of the LGBTQ.

"Me too," said Aileen and that made the ceremony even more precious.

They were both smiling and laughing.

Chapter Eleven

All the way home the two of them were acting like silly schoolgirls, singing in the car, holding hands, and pulling over at the rest stop to give each other a passionate kiss.

"We did it," said Aileen, "We are now Mrs. Parker and Mrs. Provost and are life partners. what do you think our parents will say when we tell them?"

"It will probably not be as big a surprise as we thought, they have been expecting this for a while now. Let's get back and tell Andrew and see if he has an update for us." Was Gwen's reply.

Upon their arrival at home, Aileen saw the Chief at the stable and ran over to him, yelling at the top of her voice "Chief, now what the hell is going on here?"

The Chief moved her and Gwen back out of the way, trying to calm them down, and told them. "We had a report, from an anonymous source, we believe both Stan Reynard and Aleksander Rickoff had been spotted on your property and were armed. It is necessary to respond to a threat like that with all available resources.

Sgt. Monroe and his team happened to be in Ayelsbury responding to a similar threat and responded right away. You ladies must stay back out of the way for now until we have been given the all-clear by the OPP."

Gwen countered with, "Please stop all these people from going in/out of the stable, and the flashing lights of the cars are spooking the horses. Can they turn the lights off please?"

"I will ask them to turn off their lights and try to keep access to the stable to a minimum, but the chances are slim." Responded the Chief.

Just as the Chief turned around to walk to his car, a large black SUV drove up and the chauffeur jumped out and opened the rear door. The man who exited the car was very imposing, a very large man in an OPP uniform but with significantly more decorations on his uniform and his hat than the rest. Well over 6 feet tall with a large mustache and an imposing demeanour, and he carried himself with Military bearing.

"Who is that?" Asked Aileen. "He looks a lot more important than the rest of the officers."

"He is Thomas McCarrique, the new Commissioner of the Ontario Provincial Police and the number one in command." Said the Chief. Sgt Monroe overheard their conversation and added his knowledge of the man, his rapid rise to power, booming voice, and very high-up contacts both in the police and the government.

"What the heck is he doing on our property?" asked Gwen.

"Apprehension of this criminal, one of the top 10 most wanted in Canada and Internationally, is an ongoing chronicle and his capture or extermination would be a feather in everyone's cap, especially the new Commissioner's." Answered the chauffeur. "By the way, were you aware of the reward for information leading to the capture or killing of Aleksander Rickoff?" Commissioner McCarrique interjected.

"Really?" queried Aileen. "Yes." Said the Commissioner, as he had been listening to them talk. "Are you the two ladies who have been instrumental in us having him cornered and about to be captured again after the latest debacle?"

Aileen responded a little sarcastically, "Yes, we are. Gwen, my partner, was almost raped by the man and we have been stalked, our apartment ransacked and our lives and business threatened. It appears that this man has an innate ability to escape capture, and we would not like for that to happen again. The whole town will now be aware of something going on here and we will need to explain to our parents, friends, and customers what has been happening.

What has happened to Stan, the other locally grown criminal who has been threatening us as well?"

The Commissioner called to Sgt. Monroe said, "Please tell these lovely ladies what has happened to the other criminal, Stan, and what the situation is here, while you are at it, please have all these cars turn off their flashing lights, as the ladies have requested."

Sgt. Monroe began, "Stan was wounded in the arm by larger arms fire and has been taken to the Parkerton clinic for repairs before being transported to Proctor. Rickoff has been surrounded by our officers, but this is a large search area with many places to hide, it will only be a matter of time before we have him in custody."

Suddenly, sounds of gunfire in the distance and Monroe radioed his people, "What the hell is going on Corporal?"

The response came quickly. "It wasn't us shooting. Rickoff has been spotted and fired at somebody and they returned fire. I will get back to you in a minute."

"Holy Crap." Said the corporal. We found him. He is dead. Tied to a tree by the neck. He has been shot in the chest and a single MOB shot to the back of the head."

"Who the hell shot him?" Commanded the Commissioner.

"It wasn't any of our people. It appears to have been a large caliber weapon, unlike ours." responded the Corporal.

"Stay where you are and keep signaling and we will send in the ERT team." Yelled Monroe.

The police response team recovered the body and brought it out to the driveway.

The Commissioner asked the girls jokingly, "Are you two responsible for this violence?"

"Are you crazy? We don't even own a gun," said Gwen, "We were standing right here with you, remember? He deserved what he got, for what he did to me and Aileen and for all the inhumane and vicious crimes he has committed. So, I hope he rots in hell!"

"Wow." Replied the Commissioner. "That is pretty venomous coming from such a lovely lady."

"Sorry." Gwen was sobbing, "He put us through hell and threatened our lives and our business. We are claiming the reward as well. What is it? If I am not to forward to enquire?"

"That is the next thing I will do when I return, is to put you ladies forward for the reward." The Commissioner was smiling." It's 250,000$."

Gwen fainted and Aileen caught her and they both exclaimed, "Really?"

"Yes." Said the man proudly. "And well deserved if I might add. Congratulations. Now you can go on your honeymoon, I suspect."

Gwen grabbed Aileen and whispered, "How the hell did he know about us and our honeymoon too?"

Sgt. Monroe approached the group and said to the girls, "There was an envelope nailed to the stall in the barn. We seemed to have missed it on our first search, but here it is and addressed to – *MY GIRLS* – so I suppose this is for you ladies?"

"Probably," replied Aileen as she took the envelope and passed it to Gwen. "Let's read this in private," she said.

They figured it was from Andrew but didn't want the police to see it and then take off after him as he was their saviour.

Gwen nudged Aileen and pointed to a tall man walking quickly down the side of their driveway. "That looks like Andrew, "she whispered.

The man looked over at them and put his finger to his lips, indicating **SHHH,** and then made a sign like a square with his fingers and pointed to the barn.

Aileen, being the tallest, saw his motions and secretly held up the envelope so he could see they had retrieved it, but nobody else could. He looked back, having a big smile, then gave them the thumbs-up sign, a salute, and a wave goodbye.

"Let's get up to the apartment and read the note," whispered Aileen.

"Great, "replied Gwen, "but just walk to the barn slowly so we don't draw attention to ourselves."

Once ensconced in the apartment with the door closed, Gwen opened the envelope, saw what it said, and showed it to Aileen.

My lovely young ladies, your problems have been taken care of, and you should not worry about the local stalker or Rickoff ever again. Rickoff shot your Stan in the shoulder by accident thinking it was me and then Stan returned fire and shot Rickoff in the chest. He knew I was on his trail and close by. When I confronted Rickoff, he pulled his weapon but not fast enough. I shot him in the groin to get even for all the ladies he had harmed, and then to send a message to his bosses, a single shot to the back of the head. Sorry for the mess, but Stan will be jailed for a long time and Rickoff has been dispensed with. Love you both and congrats on the wedding, so consider today my wedding present to you both.

Guess who?

BTW – you will probably never see me again unless you are in real trouble and you didn't see me this time either OK?

Chapter Twelve

The girls were ecstatic and couldn't stop smiling. After the police had gone, they went to the apartment. They spent a couple of hours to themselves, getting reacquainted as a married couple and talking through some issues that were still outstanding. They felt like new people and like a bottle that was opened with the gas being released.

Aileen said, "This is like the first day of the rest of our lives together."

It was then Gwen brought up a subject that needed to be dealt with right away, and that was the desire to get their families together and briefly explain what had been happening and the fact they were married. Aileen, although hesitant due to her mother's health, agreed and said, "Let's set a date and time and have them all together so we only have to do this once and they all get the same message. How about Tuesday afternoon at my parents' place? We can bring the wine for the celebration."

Aileen was to call her parents and see if they were available and Gwen would do the same. They needed to get Aubrey out to the stables to manage things while they were away and have him prepare for the next weekend. It's a holiday and most customers will want to go riding and deal with their charges. Preparing for this customer assault is a massive undertaking. They will need the remainder of the week to get everything in order and the place in spit spot shape and will also see if Abigail could spare some time as well.

Gwen was thinking, "I think we should have the Loughty parents there as well as the twins are being affected and it will more than likely get worse as your mom gets sicker. What do you think?"

"Great idea." Replied Aileen. "I will call them right away as well. "

Aileen opened up a conversation with Gwen saying, "We need to start thinking about what we want to do with the reward money and brainstorm some business and personal ideas."

"Yes, we do." Replied Gwen, "I have a few ideas myself.

"What do you suggest?" queried Aileen. "We need to think about it and try to develop a priority listing, but first, we need to address the family conversations."

It was 2 p.m. at the Parker residence and both the Parker and Provost families were there. Aileen took this time to explain to the two groups they had invited the Loughty parents as well. The incidents at the stable and the potential fallout will probably affect them as well, with Abigail at the store and Aubrey working at the stables.

The Loughty family turned up and seemed a little confused. Gwen and Aileen proceeded to open the wine and set out the glasses, hoping that this might mellow the responses.

It was decided that Aileen would begin the session by introducing Grayson and Amelia Loughty to everyone and begin to make her lengthy presentation.

She began her explanation with, "We have been dealing with a lot of issues lately, and rather than get everyone involved and worried, we decided, as adults and business owners, to handle things ourselves as best as we could with the police involvement."

She elaborated on her extensive time with the Council, causing issues between her darling Gwen and herself and this precipitated Gwen going to the show in Proctor alone. "After the perpetrator attempted to rape her and the Angel saved her, everything started to go even worse, if that was possible. We worked out our issues and are back together and happier."

Gwen took over and went over the attack, the existence of "Angel," the stalking by Rickoff and Stan, and the horrible situation at the stables involving the Police and the OPP. The appearance of the powerful OPP Commissioner was a bonus to address the shooting of Stan and the death of Rickoff. Rickoff was an international drug dealer and human trafficker, and on Canada and the US's most wanted list for 2 years. She explained how it was he escaped custody twice and came looking for her as she could identify him."

The entire six-person audience was stunned and wanted to ask questions, but Gwen just asked them to wait until they were done.

Aileen came back to take over as Gwen had started sobbing and needed a break.

She gave some background on the stalking, the emergency call to the 'Angel," the break-in at the apartment, and the fact that all this turmoil was affecting their safety, relationships, and customers. "All is almost back to normal now and we are concentrating on the business success and our relationship."

They would ask Aubrey to spend more time with them if possible and look forward to doing a couple of shows in the fall.

"Any questions?" She asked.

James was the first, "Is everyone OK now, and nobody was hurt?"

Gwen replied, "Aside from the attack on me and dispatching the two criminals, all is fine."

Matthew was next, "Who the hell is this "Angel" person"?

Aileen answered, "Sorry, but we are not at liberty to discuss him with anyone. This came from the Commissioner and higher. We can tell you that he was instrumental in saving Gwen from a far worse attack and came to our rescue twice when we called his special number. "

Matthew replied, "This sounds like a spy story or TV police programme."

"His life would be in danger if people knew his real identity." Answered Aileen.

Amelia interjected, "I had the two victims at the clinic and Stan was shot in the shoulder and will be ok, aside from going to jail for a long time, but the other man, Rickoff, was shot three times by someone who was very angry and died instantly after the third shot. The bullets used to shoot Stan and the ones that killed Rickoff were from different weapons."

Amy spoke up with very little energy, "Was it the police who shot these people, or did you ladies shoot them?"

Gwen answered, "No. We didn't shoot them, although we would have if we had the chance. That is all we can say other than to let you know all is well, the business is doing fine and we are happy again."

Aileen looked at Gwen and nodded; It's time now.

The two of them stood up and declared, "There is one positive thing that happened and we want to tell all of you. They held hands and announced together. "

"WE ARE MARRIED,"

This proclamation brought a solemn hush over the group aside from Amy who struggled to speak but said, "Congratulations, I am very pleased and happy."

James was confused, "But how and when, you have been too busy?"

"Last week at the JP office in Ayelsbury." Gwen smiled.

James again," Will there be a regular ceremony sometime?"

Gwen stood up, "We were angry as the priest and reverend in Parkerton who refused to perform the ceremony, so we went to where they would gladly help us. We would like to have a small reception soon if that is ok with all of you. We just need time to get things around the stables up to our business standards and then we can plan a small reception."

Aileen and Gwen stood up and asked, "Are there any objections?"

All the heads nodded no and actually, Louise giggled, "Maybe I will get grandchildren after all."

They all spent the next hour chatting. The girls made their apologies, saying they were still on their honeymoon, heading back to their apartment to put things back in order.

Once the girls had departed, the parents and concerned party were left there with astonished looks on their faces and a million questions they never thought to ask at the time.

"So, what do you make of all that?" James enquired.

Matthew was the next to speak, "I think those two ladies are the bravest people I have met in a long time. To deal with an international criminal, Stan from town, people telling them they are not allowed to talk about things, and especially their secret marriage in a town like Parkerton is amazing."

Amelia injected a few words, "I have seen those two women in the clinic on several occasions, especially the issue over the drugs, and I couldn't be prouder if they were my own. The positive influence they have had on Abby and Aubrey has been a blessing.

Just look how they handled the drug issues, murder, and coming out at the school, not even counting the graduation."

"Louise, what are you thinking?" Asked Matthew, her husband.

She responded sheepishly. "All this has been going on in our daughter's life and we have been practically oblivious because we have been so busy with the new businesses.

Thank goodness she had Aileen and all of you to be there for them. I promise to be a better communicator and mother from here on."

Amy spoke very softly as she was beginning to tire quicker every week, "Louise, please don't blame yourself, as we have all been very busy. We are blessed with having two strong and determined women in our family and I would also love to have a grandchild before I die."

Amy continued, "All of you may/may not know, but I am dying from ALS. There is no cure and the disease ravage the body, so I have been less involved in their lives as well. I wish to thank Amelia and Grayson as the twins stepping in to help at the store and the stable has been a godsend. Thank you."

The ladies were crying and the men were choking up. This was a brave thing for Amy to admit to people, but knowing this would allow everyone to be there for her as the condition worsened.

James stood up and said, "Thanks everyone, but I need to get Amy up to bed and take her meds. We must try to get together more often to support the girls, Amy, and everything else."

There was no doubt that when the Provost and Loughty families returned home, there would be some interesting conversations regarding the news that they were given by the girls and especially the secret marriage.

The girls left and went back to their apartment to discuss their future, not even knowing what was about to come their way.

Chapter Thirteen

James had called the ALS Society and requested their assistance. Would they come and speak with him and take a look around the house? His wife is suffering from ALS, and he is worried about the future and how to address her upcoming needs.

Just as he was expecting them to arrive, the next day, his 3rd in command, Peter Underhay, came to his door and asked if he could speak with him.

"Come in Peter, " James invited. "What can I do for you?"

Peter put forward a suggestion. "I spoke to my wife about you needing someone at the dress shop and now the kids are all in school; she would like to volunteer. She had 10 years of experience in women's retail and ended up being the manager of the store just before we left to move here. "

James was delighted as Peter Underhay was one of his best managers and organizers. "When can she start?"

"As soon as you need her, "Peter replied, smiling.

"How about tomorrow?" James was desperate.

"Done." Said Peter.

James just told him there was a young girl at the store named Abigail who has been trying to keep things afloat but is struggling and I will tell her your wife is coming. "BTW, what is your wife's name?"

"Glenda." He replied.

Glenda was a petite lady in her early 50s with a hint of grey hair who had been active all her life; with the kids grown and having lives of their own, she was beginning to get restless and looking for something positive to do in town. When she was approached by her husband regarding the dress shop, she jumped at the chance, as she had spent 10 years in retail over her life, most of it in women's clothing. This was a great opportunity, she thought, and a chance to help out the Parker family.

Glenda went to the store the next morning and introduced herself to Abigail. "I hope I am not stepping on your toes," she said to Abigail.

Abigail replied, "No, you are most welcome, as I have been struggling and we can use your experience and knowledge to do the best job we can and make Amy proud."

In the Phoenix building at that exact moment, the intercom rang in James's office and it was Jocelyn, James's amazing assistant. "The ALS people are here to speak with you. Can I send them in?"

"Yes please." Finally, he felt like he was making some progress.

The ALS people outlined all the services and consultations they provide along with designs for home access and especially for bathroom access.

They also can provide financial assistance should it be necessary, as some of the much-needed appliances, for mobility, can be pricey.

They tried to soften the prognosis to not get him too worried about the immediate implications of the disease, but James had done his research and was well aware of what was to come but was oblivious when it came to mobility issues for Amy.

The ladies answered his questions but reminded him that knowing the potential of what was to come and experiencing the effects of the disease on a loved one are exponentially different situations. This will be hard for you and the family.

The lead lady expressed a reflection. "James, we will be here for you throughout this process, but keep in mind that this is a terminal diagnosis, and you must be prepared for the end as well as the trip. We will put you in touch with some local people who have made this journey if you would like to talk to someone with experience?"

"That would be super." James was on the verge of crying.

The leader reiterated that it is also a key piece of this process that you have a serious discussion with your daughter and friends to make them aware of what is/will happen.

"Will do." Replied James. "Aileen is an intelligent and strong woman who is going through her struggles right now, but we are very close and she knows what is happening."

James struggled with the next question, "My wife has expressed her desire to not be allowed to be in major pain or suffer from immobility and be a burden to the family. She has been reading about the MAID, Medical Assistance in Dying, services available in Canada and wanted me to investigate."

The leader answered, "This is not an uncommon request from the patient, but it might be a little premature to discuss that option, although it may come to that later.

We will send you all the information we have available. Sorry, but we have another patient to visit in Ayelsbury, so we must get back soon. Can we see your home?"

"I will take you right now and you can follow me so you can leave directly after. It is only a couple of minutes from here." James told them.

"Jocelyn, will you hold the fort until I return." James requested.

Once they arrived and walked through James's new home and talked briefly with Amy, they had a myriad of suggestions for him and the residence.

The leader said briefly, "We will send you more information now that we have seen your home, but you need to begin thinking about interior and exterior mobility access for Amy around the home. Walkers, a wheelchair, grab bars for the bath and shower, and especially more accessibility to the bathroom and shower.

"The installation of a wheelchair ramp is critical and you need to consider making space for Amy on the ground floor to eliminate the stairs and allow her some freedom. I know this sounds daunting, but if you can get priorities done NOW, like wider bathroom access on the main floor, purchase/rental of a wheelchair, the front ramp, and the full-time services of a PSW, it will make things easier."

"None of this is mandatory for you to do, but if you wish to make the time Amy has remaining a lot more comfortable and life for you and your daughter less worrisome, these arrangements are a priority."

James thought for a second. "I will get one of my Engineers working on the house issues right away and will challenge my daughter to contact you and acquire the walker, wheelchair, and immediate accessories required for the bathroom and shower."

Later that afternoon, he gathered Amy, Aileen, and Gwen together to explain the situation and outline his short and longer-term plans. He invited Gwen as he knew she was an integral part of Aileen's life and could be a great asset.

Amy was concerned about the state of the store and her clients. James was quick to reply, "We have found a person within our organization who has a substantial background in retail. Her name is Glenda and she can start tomorrow. She can work with Abigail and keep you informed weekly about their progress and sales."

Amy asked James, "Can I meet with her to discuss some of the procedures and processes we use as well as ordering? It will be necessary for someone to have signing authority for cheques and orders."

"Will bring her around in the next few days. OK, he answered. Her name is Glenda remember?"

James was thinking that it worked once, so why not try to petition the workforce and see if there might be somebody in one of the families with a nursing or PSW background who would be willing to work for him as required to assist Amy?

Chapter Fourteen

He called Grayson and Jocelyn at the office and asked them, "Please spread the word to the troops and their families to see if there might be a resource close by who has a PSW or Nursing background. Grayson, can you speak with Amelia and see if she is aware of anyone who might fit the bill for us through her contacts at the clinic? Then I need to speak to you, Grayson, regarding the issues with the house. How about tomorrow morning?"

"OK boss no problem. I will meet you at the house at 8 am. OK?" Grayson was curious.

"See you then," said James.

James was trying to get his head around all the construction and changes to be made to their home and was seeking advice from Grayson as to how to proceed. Bad news and good news was coming from Grayson and Jocelyn.

Jocelyn called back in one hour, "We have located one of the ladies who nursed for 25 years and is willing to speak with you about Amy's needs and your expectations. Her name is Jo-Anne and she is the wife of your brewery manager. Let me know when you want to meet with her, boss."

"How about tomorrow afternoon," James was thrilled.

His meeting with Jo-Anne the next afternoon went very well; her personality and experience as well as the fact she had specialized in assisting these types of patients, were just what he was searching for, so he made her an offer. She accepted right away and they went to meet Amy that afternoon.

Amy was thrilled that she would have someone with experience there for a few hours in the morning as well as a few at dinner and bedtime, so the deal was done and Jo-Anne started the next morning. Amy was feeling very upset about the extra burden being put on James and she kept researching the MAID solution, much to the chagrin of the family.

Jo-Anne and James agreed that they would ask Amelia at the clinic if Amy could stay at the clinic for a few hours each day under the supervision of Jo-Anne until the renovations to their house were completed and she could be moved in full time. James would call Amelia in the morning.

James received a letter from the company that dumped them and moved on. They were threatening to reopen the agreement. They were aware of the major changes James and his teams had accomplished as well as the sale of company houses and wanted back in.

James decided to do nothing and just ignore their plea but to contact the company lawyer and make them aware of the situation and get their advice.

When James and Grayson had discussed the requirements and suggestions to make Amy mobile and comfortable, Grayson spoke to

James, "Even if today we started, the disruption, noise, and mess would be unbearable for her, and it might keep up for 2-3 months. So, we have a couple of options:

1. We do the absolute priority items to enhance the quality of life for Amy and fix the bathroom access, main ramp, and handrails as quickly as possible.
2. She could stay with someone for a few months until we were done but would face the same access issues there as well.
3. We could retrofit the empty company house, which is identical to yours already, and get the project done much quicker. When we move, some of the visible change would be a lot less of a shock.
4. The priority, regardless, is to install the ramp and the handrails in the bathroom and widen the doorway while we do the other house.
5. Once you no longer require the other house, sorry for this. We could use it as a retreat for ALS and handicapped persons and get the province to put in money for expansion."

James asked Grayson, "Which do YOU suggest?"

"We do #1 and #4 together and start quickly on your existing house to do the priority items. Then, we get a local contractor to put in a quote for doing the other house. We need to get a contractor right away to do the bathroom and the ramp. You handle the old company and the new budgets and I will deal with the ramp and bathroom. Ok"

The next morning, James called a contractor who had done work for the business and had him come to talk about the immediate installation of the two priority items and give him an estimated cost. James was getting confused about what to do in the short and medium term as Amy seemed to be getting worse every week. He called the Dr. and the ALS people for advice on house renovations and the best way to go.

Unfortunately, the Dr. and the ALS people had the same advice and this scared him even more. They both inferred that doing major renovations to the existing home or taking over the other house and retrofitting the entire property might be the wrong thing to do, depending on her condition, as her time might be shorter than the renovations would take.

They both recommended that he concentrate his efforts and money on preparing the first floor of the current home by adding a large bedroom, replacing the whole bathroom, installing a walk-in tub, and especially getting the ramp at the front installed.

This would mean going to the town to get permission and building permits, so he better get the process started. The fact that they were such a prominent family in town and were dealing with a terminal illness should expedite the process considerably. It was time to use his status in the town to his advantage for a change.

James had taken the suggestions, although depressing, from the Dr. and the ALS people about getting a new house in town and doing all the construction work in preparation for Amy's arrival. Both parties agreed that not knowing the length of time Amy had remaining vs. the reworking of the new house just didn't make for a good outcome.

Their shared suggestions were to get the existing home up to a standard to make Amy as comfortable as possible without worrying about the timing and cost of the new house. Giving Amy the most comfortable and accessible environment as quickly as possible was a priority.

James contacted Grayson and he arranged for a contractor to get started on the ramp and the preparations on the ground floor for a larger bedroom, walk-in tub, enlarging the whole bathroom to allow for wheelchair access, and lowering the sink was the priority.

Grayson would search for a contractor to get them started putting together a plan for the house. James would go in person to the building permit office and arrange for the necessary permits and expedite the process as time was at a premium.

The building permits were ready the next day, a boon when dealing with small-town service, and the contractor was at James' office at 10 a.m. the same day. James outlined what he needed to be done, the timing for completion, and also to minimize the potential for noise.

James had decided to find an alternate location for Amy to spend the day while the construction was underway. He just had to figure out where she could stay all day and be safe. He agreed that staying at the clinic, as long as she could be a short-term answer.

The contractor said his demands were perfectly doable; they would need a week, and he would put his entire crew on the job as this was a medical priority.

Another note from the company in Proctor demanding a meeting, and this time, James decided to see what they wanted but not before calling all his Senior people together and explaining the situation as he saw it.

With Amy's current terminal situation, the issues with the store, and the girls, he was in no frame of mind to deal with this nonsense. He called Grayson and Jocelyn, requesting they arrange for all Senior Managers to meet tomorrow at 9 a.m. to bring them up to speed.

James called the person in Proctor and said he was arranging a meeting for them on Wednesday that week at 10 a.m. and not to bring more than three people. If this is not suitable for your schedule, then the whole thing is canceled and we will see you in court. He told Grayson what he had told them and Grayson was in full support.

The meeting of the Senior people began with James giving a short statement of the current situation and what the people in Proctor were trying to accomplish.

"Gents, welcome to this update and I hope we will all be on the same page. The CEO at ErgoPoly in Proctor reviewed our situation 3 years ago and was not happy about our performance and the costs related to the housing and terminal lack of work. We, as Phoenix Resources, proposed to take over the facility and swallow all the personnel costs, BUT we would own the building, equipment, and all properties associated with the location, and they agreed. Now they have seen what we, as a team, have been able to accomplish with trust and hard work; they want to revisit our agreement as they say the CEO was not medically fit to make that decision.

"I have told them to go to hell and our Phoenix team is happy with the way things are going and has no desire to change except for expansion. They want a meeting anyway, so I have scheduled one for tomorrow at 10 a.m. here and will tell them the same thing. If there is anyone in this room that disagrees with this decision, please raise your hand."

There was no pause, and no hands went up.

"Great, no hands are showing. Once we finish with this foolishness, we will have a General Meeting for the troops and discuss any outstanding issues. In the interim, please get your supervisors to speak with their teams to get any concerns and/or suggestions in writing, and we will address them all. Thanks for coming, people, and keep up the exceptional standards. You are my heroes."

The meeting took place at the Phoenix Resources office and three people from the ErgoPoly group attended as well as James, Grayson, Peter Underhay, and the Phoenix lawyer, Jocelyn, who would be taking notes.

The ErgoPoly people began with a very aggressive stance, indicating it was their position that the CEO at the time had no responsibility or power for making the decision he did, and they had every intention of reversing the decision and returning to being major stakeholders in the new company.

The Phoenix lawyer met this indicating, "Gentlemen, this decision to separate and to divest ErgoPoly of all responsibility was made three years ago, and since that time, the people at Phoenix have lived up to their part of the bargain. They have assumed the outstanding debts, completed the existing contracts, signed extensions, sold off the company homes, and transferred the costs to the new owners. We believe that Phoenix Resources has surpassed its agreement with ErgoPoly and as such has no interest in revisiting the agreement. Also, **WE** believe it is only due to the financial successes and expansions of Phoenix and the embarrassment this is causing ErgoPoly and its shareholders that brought you here today.

"We see no need to go further with this meeting and, if need be, will see you in court. We will also be compiling an article for the local and provincial newspapers outlining your ludicrous claims and the fact that as major players in Parkerton and the County, we have an obligation to our owners and citizens to look after their interests first. Having your group back involved does not conform to our corporate guidelines and/or Mission Statement. We are calling a close to this meeting and wish you safe travels."

At this juncture, the Phoenix personnel and lawyer rose to leave, but the ErgoPoly people began yelling at them saying, "You have to hear us out and understand our position."

James interjected, "No, we do not, as we know your position. The company in Proctor is in trouble and losing money, causing their shareholder's panic. You want to move some offices back here, hire employees from Proctor offices that you are about to lay off and share in our successes and plans. The answer, gentlemen, is a resounding **NO.** So, Jocelyn will show you out."

The ErgoPoly people were livid. It was obvious by their reaction that they knew they were beaten and had to return to Proctor with their collective tails between their legs and report to their Head Office and shareholders.

After they had departed, the Phoenix group was cheering and shaking hands. They knew that Jocelyn would be spreading the word through the company grapevine and would await the reaction.

James said to Jocelyn, "Feel free to post these minutes on the company notice boards, with a note at the bottom telling people to contact me directly should they have any issue with the meeting resolution."

Chapter Fifteen

The entire Loughty family had been so busy since they left the Harrelson farm, where they had been squatting for three months, and the place they had discovered Aileen in the bush after her failed attempt at running away from home, that they had never thought about checking their mail. The Harrelson estate constituted a 150-year-old home, shed, and large barn, along with 200 acres of undeveloped land. From all the past gossip and police reports, the property was used for the manufacture and distribution of alcohol during prohibition, hence the large vats in the basement of the barn, large storage shed, and underground cooling cave.

The Loughtys believed there was no rush to get the mail, as nobody back in the UK knew where they were, and people in Parkerton usually communicated by talking as opposed to using mail. The only mail they believed to be there would be any old bills at the house and the utility bills at their new place. They have been paid automatically from the bank since the introduction of the latest banking procedures, so no rush.

Abigail went to retrieve the mail. They were correct and 90% of the mail was either junk or bills, but there were two large envelopes with none of the usual addresses, so she was curious and couldn't wait to get back home to open them.

When she unsealed the first envelope, she found a single sheet of paper with the following note appearing to be made from words cut from notes, papers, and books. The note read:

We want the box from the farm and know you have it. will be in touch.

This scared Abigail and she called her father right away. "I will be there in five minutes, don't panic." He was very anxious. She was going to open the second one but was very troubled so she waited. Five minutes later, her father came through the door and looked at the note. "What the hell are they talking about?" he muttered. "Give me the other one and sit down over there, please, Abby." He was upset.

As he opened the second one, he could see it looked the same as the first, but the wording was longer and more severe.

We are not kidding. Leave the box on your doorstep tomorrow and we will pick it up. This is not a joke and if the box is not on the step, there will be severe consequences. If there are problems, call 905-555-6666 Saul.

"I think it is time for us to call the Chief." Abigail was stammering.

'Yes." Her father replied. "This has something to do with the barn at the Harrelson farm and all those criminals who were after this 'box'. "

"Do you know what they mean about the box?" Abby asked.

"No, but these must be some of the same people who were here before and are very dangerous. We need to involve the police. How many other things can go wrong in Parkerton, as it is like living in a TV movie."

"I remember the fuss that was made over the box, so there must be something very valuable to someone inside. The Chief told us these people are seriously dangerous and believe he said they were associated with a mob in the city." Grayson's voice was shaking.

When they called the station, they asked for the Chief and he told them it was Grayson Loughty and it was urgent.

"Chief Collard here, Grayson, what can I do for you, and what is so urgent?"

"You need to come to the new house, please," Grayson pleaded. "We received a threatening note from the gangsters who were at the farmhouse. You remember that incident, right?"

"Yes, I remember it well. "Chief replied. "Who could forget those bloody criminals."

"Hurry please." Grayson pleaded again. "They are demanding we give them the box from the farm or else. I have no idea what the hell or else means, or what they are talking about."

The Chief arrived and looked at the note. "I don't know what they are referring to either." He spoke. "Maybe we need to put together another search party and go back to do a more thorough job. We have GPR and metal detectors to help. I will get everyone there at 10 a.m. Tomorrow we will find this damn box, once and for all. Don't get all panicky, as I will have an officer drive by every hour until morning. See you at the barn in the morning."

The Chief was required to check with the property owners, which currently was the local bank, before doing any work or investigation. The bank manager informed him that James Parker at Phoenix Resources is currently in the process of acquiring the property and it looks like it is a done deal today.

Please check with him, and this would be James Parker, the CEO.

The Chief called James, explained the situation, and asked for permission to do the extended search and he said to go right ahead.

James explained to the Chief that they were trying to keep his acquisition under wraps until it was completed, but now the cat is out of the bag, we have no problem. Remember to keep any damage to a minimum and whatever you find belongs to the company. The Chief gave everyone the clearance and they were to be there in the morning to start the search.

At 10 a.m., everyone was assembled at the farm, and the intense search began. "Don't leave one stone unturned or one corner not searched as this is a very serious matter." Yelled the Chief.

Grayson, as second in command, forgot about the Phoenix Board's decision to purchase the property and felt a little better than the weight was now on the shoulders of the company and not him and his family. He needed to speak with James as soon as possible.

The search had been happening for over 2 hours, and they had found nothing; then one of the officers shouted, "I found a hidden room."

The room was well-hidden behind old workbenches and covered with straw. Once they removed all the detritus from in front, they pried the door open, and they saw a large item wrapped in a tarpaulin, covered with straw, and sitting up off the ground.

The officers dragged the article out into the open, removed the straw and the tarp, and there it was 'The Box. '

"Bring it out where we can see it," yelled the Chief, as he was anxious to finally find out what all the fuss was about. "I think we should take it back to the station to open it as we never know what they might have installed as hazards or tricks."

Once they were back at the station, James, Grayson, and the Chief took the box into one of the cells as protection, just in case, and the Chief called to his officer and said, "Bring me a plastic or wooden ruler."

The officer found an old wooden ruler and the Chief proceeded to open the top just an inch or so and slid the ruler along the rim with just a few inches of it inside. "Looks ok," he said. He opened the top the rest of the way and the three of them stood with their eyes wide and mouths open, not believing what they were seeing inside.

"Is that what I think it is?" questioned James. Grayson, speechless, finally replied, "I can't believe what I am seeing, but it looks like money and gold."

The Chief picked up the paper and looked at it, "These are Bearer Bonds from the 1940s. There are 6 gold bricks, plates to print money, and a small wooden case. When he opened the case, the contents were even more startling, as the light was sparkling off the contents. "These look like diamonds," He stammered. "And there are a lot of them. This is like a pirate treasure and must be worth over a million dollars, even more if the Bearer Bonds are still cashable. "

James and Grayson were stunned.

"What the hell do we do with all this Chief?" asked James.

"Damned if I know. But we should probably contact the OPP and RCMP as quickly as possible." He replied.

It took two days for the OPP and RCMP to gather the people they required and travel to Parkerton police station. Once both police departments had arrived and viewed the contents of the box and to figure out where they all belonged, they told the Chief, James, and Grayson they would have to catalogue all the contents and take them back to the professionals. Once this was completed, they would get back to them. Grayson said, "What about these people coming to my house tomorrow?" After about an hour of discussions, the officers contrived a plan to attempt to capture the perpetrators and requested assistance from Grayson and James.

Sgt. Monroe of the OPP, the officer initially involved in the case and who had followed along with the last incidents, suggested the course of action.

"We need to get Grayson to call the number they gave you and ask for the person they named." He outlined. "Tell them you found the box in a secret compartment at the back of the garage and are going to leave it for him in the same place so nobody can see it from the road. Also, tell them they have 24 hours to come and get it, or you will be taking it to the local police. Don't say anything else or get into a conversation."

He continued, "Chief, we need to remove some material from inside the box and just leave the minimum in case they escape. " The Chief said, "Please give the box to Grayson and allow him to take it to the barn alone just in case someone might be watching. If there is no one around we will go in and install some CCTV cameras at select locations and get places for the constables to settle and prepare to call for the troops if/when these people arrive. Any questions from anyone?"

The next morning arrived.

Everything was in place; the box was re-hidden, cameras installed, and constables hidden in the surrounding shrubbery, so now they just waited.

At almost the end of the identified timeline, a vehicle drove up and stopped at the driveway entrance. It was a larger pickup truck carrying four nasty-looking persons. They proceeded to back into the drive, prepared for a quick retreat, and left two out front to guard the area and the truck.

A shout came from inside, telling the guards, "We found the box." The obvious leader shouted back, "Bring it out here to the truck, and don't open it." Once it was in the back of the truck, secured and covered, the Chief hollered into his radio, " Now Go."

All the constables from the OPP, RCMP, and local station, directed by the Chief, converged on the vehicle and arrested the perpetrators without incident. They were armed but had no chance to draw their weapons before they were arrested. "A great collar." Sounded the Chief, with a huge smile on his face. He directed everyone, "Let's get these clowns and the box back to the station and turn them over to the OPP or RCMP for processing.

The Chief asked for everyone's attention and had some comments. He started," As for the contents of the box, we will add it to the material we already have and let the RCMP sort out where it goes. It is important to remember there is a new owner of the property, and their rights to whatever is found on their land need to be a priority."

They all left, perpetrators in handcuffs inside the police vehicle, and headed to the Parkerton station.

Once they arrived at the station and had the people in the cells, the various police entities were gathered in the conference room, trying to sort out the contents of the box.

The RCMP were adamant that the whiskey, printing plates, and old bearer bonds as well as the diamonds, needed to go with them to ensure they were not stolen goods. The gold could stay for now, but photos were taken of each bar to double-check their validity before being given to James and Phoenix Resources as the property owners.

Chapter Sixteen

Aileen and Gwen were still brainstorming ideas and did not know what was transpiring at the Harrelson barn.

Gwen was spewing forth ideas. "How about a canvas/plastic cover for the paddock area to expand the stable area and accommodate more horses? Move into my parent's house that is sitting there doing nothing right now, open a new business – like a small nursery -breed Artemis and Bluster with a Trakehner stud, and bring Aubrey on board full-time if he wants."

"Wow," Aileen was excited. "They are super ideas and you must have been thinking about this for a while now. Anything else?"

"Yes, but you might not like this additional one," Gwen replied timidly.

"Try me." Aileen was curious.

"I have been thinking about us having a baby." Gwen was sobbing.

"So was I," answered Aileen, "but this would mean some drastic lifestyle changes. Let's sit down tonight and figure out our priorities and where we go from here. "

"OK," Gwen was enthusiastic.

"So, let's list the priorities and get started." Aileen prompted. "So, what do you think about that list my darling?" asked Aileen.

“I like it, so let's identify who does what and put the initial at the front of the list.”

“OK,” replied Gwen.

- **G&A:** Locate a local sperm donor clinic for us and determine which of us will get pregnant first.
- **G:** Speak with Gwen’s parents about moving into the house ASAP.
- **G&A:** Explain to parents what we are doing!!!!!!
- **G&A:** Rearrange the paddock and exercise area, leaving the eastern area open for a new business. Contact the same local contractor for assistance.
- Get quotes for the installation of the paddock and exercise area covers.
- **G&A:** Expand the stable area for more residents, using the same contractor.
- **G:** Check with Aubrey about the full-time position.
- **A:** Locate semen donor for horses.

“I have an idea as to which of us should get pregnant first and I believe it should be you, sweetheart, and have the baby while your mother is still with us. What do you think?” Gwen was crying.

Aileen answered, "No, I think it should be you. Your mother was the one who mentioned grandchildren and we have no idea how long my mother is going to be with us. She is going downhill faster than expected and is in more pain every day. Dad and I need to discuss the next steps. This will probably mean that I will have to be spending some more time with both of them off and on and besides that, you are more familiar with the stable operation than I am and know who can deal with the additions and expenses."

"OK," replied Gwen, "but we have some major decisions to make before we ever contemplate having kids. Who is going to research human semen? Are we financially and mentally prepared for this large move?"

"The first thing we need to do is to call our Dr. and arrange for a meeting to discuss this whole process. I understand there are many safe ways to accomplish this result, especially ruling out the normal way for a woman to get pregnant."

Gwen was looking very sheepish and didn't want to disclose her secret right yet as she was sure Aileen would be angry and disappointed, so she hid the fact that she had already been to the Dr. and taken matters into her own hands. She had been feeling a little off and was sick in the morning, so she was almost sure she was already pregnant.

"We need to decide on whether or not we are going to the show in Palisade in two weeks, and when do we plan to tell our families again of all the new stuff happening and the potential of becoming grandparents?

There are dozens of things we need to consider, but we have a list and should follow it." Answered Aileen.

"I have spoken with my parents and we can move into the house anytime we wish and get prepared for winter arrival, so that might be the first step." Gwen was excited. "There are so many large things to do Aileen and I am so thrilled." "Great, " said Aileen. "Let's get going with moving into the house, and we can decide on the rest later. I think we should do the show in Palisade as it might be the last time we ride for a while, and there will be vendors there to assist us with the paddock design."

"Let's go across now and see what is in the house and what we need to bring with us. Ok?" Gwen responded excitedly.

"We will probably just need all our stuff and some bedding as the house has three bedrooms, all pretty much furnished, one for each of us and one for a nursery. Ha ha.

The appliances are there and it will be nice to have a real non-microwaved dinner for a change. There is plenty of water and heat, lots of room." Gwen was smiling.

"Let's go, "exclaimed Aileen, "this is exciting and a whole new chapter in our lives and new family."

They had moved into the house and arranged all their belongings, which didn't amount to much, and were planning their trip to the next show. They needed to make motel arrangements this time around to eliminate the possibility of anything happening to them, especially Gwen. Gwen was very apprehensive about going to another show after the incident in Proctor and the resulting disaster to their lives and happiness. She was worried about a show jumping if she was carrying a baby.

It had been a couple of months now and they were just getting settled; they required Aubrey to handle the business and get the truck and the trailer all packed. Bluster and Artemis had not been exercised or ridden for a while and they decided to take the two of them for a run and a turn around the practice ring.

The show was in five days and they were still preparing.

Aileen asked, "Sweetheart, can we talk please? I have been thinking a lot about all the things we have going on as well as the health of Amy and my conclusion is that we review our decision to be parents so soon and put it off until next year. What do you think?"

Gwen was very slow to respond and looked like she was crying. "I understand what you are saying and don't disagree". She was almost in tears, keeping the secret was killing her.

"Maybe we became too excited by the potential of being pregnant and being parents. Once we get the covered stable sorted out, both of our charges inseminated and foals on the way, one or two last shows, and the business on firmer ground, we can look at the possibility."

I would like to present a baby to your mother while she is here and to mine just to shut her up, but we need to be doing this for us and nobody else. Maybe we even need to get some permission from Andrew, she said laughingly."

The girls had been busy preparing for the show on the weekend and forgot about needing the horses to have shoes before they left. They called the farrier and explained their dilemma and asked if he could come out within the next day or so. Mr. Baldwin, the farrier, would be there the next afternoon and would look at the hooves of all the others while he was there as a precautionary measure.

Gwen was having second thoughts after the incident at the last show but was encouraged that Aileen would be there. A motel room was booked and there were many places to eat away from the exhibition grounds. So once the shoes were on, Aubrey was on board with looking after the place and Abigail was coming to help when she could, and all the feed and tack were looked after, she felt better about going.

When Friday came, Gwen and Aileen called their parents, informed them of their decision and their plans, and said they would call daily. They loaded up the horses, tack, and hay and were all set to hit the trail.

They were just getting ready to leave when Aileen was doing a final check of the stable and found an envelope tacked to its normal place.

She hurried to show Gwen and said, "Sweetheart, look what I found in the stable."

An envelope giggled Gwen, "I wonder who that might be from?"

Aileen opened the note and read the content.

To my special girls. Have a great trip, and do well. I will be around and watching, so nothing will happen.

Love You Both ---- Andrew

Aileen laughed, “It is like having our special guardian, Angel.”

Gwen snickered, “Now I feel much better so let's get going, OK?”

Chapter Seventeen

They arrived at the fairgrounds, installed the horses in place, arranged the tack, and headed for the motel.

The next morning, they did a practice ride and their charges were calm and well-behaved.

After the first two rides, they came away with two second-place finishes but when the judge announced the arrival of Aileen and Gwen from Phoenix Equestrian they had their lavender jackets, and the Phoenix blankets were on full display they were very proud.

Day two started the same way and all was well but during their second set of rides Bluster threw a shoe and couldn't finish safely, so, Artemis had to carry the load.

She proved to be up to the challenge and finished in first place.

They returned to the stable and wiped down the horses, treated them with some apples and oats, and headed for a bite to eat before they loaded up to leave.

As they were walking to the truck, Aileen's phone rang; it was Aubrey. Aileen put him on speaker.

He started with, "I wanted to wait until you had finished your rides before I called. There has been an incident at the barn. No one was hurt and no major damage was done, but there has been malicious graffiti scrawled across the barn doors and siding gates.

It also looks like someone had tried to break in and do whatever damage they could. You don't need to rush back as I called the police and your parents and all is under control. See you in a couple of hours."

Gwen was shaken, "Let's just head back and we can grab a bite along the way."

Aileen was already turning to head for the truck and start loading.

45 minutes later, they were on the road and called Aubrey to let him know.

"There always had to be something, didn't there my love?" Gwen was sobbing.

"We will be there in less than two hours, so stay calm and remember who is watching over us, Aileen said, trying to stay calm. Also, the CCTV cameras and the motion sensors we had installed should have captured these assholes and we will prosecute them."

Chef Collard called the girls and their parents to come to the station to watch the video and see if they recognized the perpetrators before he began an all-out community search.

"Does anyone recognize these idiots?" questioned the Chief.

Everyone assembled repeated the same sentiment and their answers were NO, they don't look familiar at all.

Aileen interjected, "They look a lot like some of the teenagers from the high school and you may want to start there."

Gwen was angry, "Who the hell would be stupid enough to do this, and what have we done to deserve this type of treatment?"

James replied, "Remember that since you approached the local churches to handle the ceremony and they all refused, the churches have been preaching at length about following the word and a marriage should be between one man and one woman. This quite possibly has stirred up the less open-minded to become radicalized and want to take some action."

"Does this mean we are not safe?" asked Aileen to the Chief.

"I can't say for certain, yes or no, but we will track down these clowns and make sure they are severely punished and made an example for the community." Responded the Chief. "Just be cautious for the next while and make sure all the doors and barn access points are locked."

"But that lunatic had a gasoline canister and do you realize the damage and destruction that would be the result of a fire, with all that wooden structure, the hay and straw laying about, and nobody there to use the fire extinguishers?

We have a dozen horses, some belonging to our customers and if word gets out that we are potential targets and their animals might be compromised, people will begin to move their animals to another venue and we would be ruined. You need to find these people quickly before our guardian angel does and he sorts them out."

"We will be back at the stable if you need us for anything." Aileen offered. "We have a lot of issues that need to be addressed. The weekend is coming and some of our customers will want to get out on the trails for the first time in a few weeks and we need to be ready for them."

They were back at the stable and attempting to do all the outstanding items in a few hours. It was extremely hot and humid in the barn and the girls were perspiring profusely but had neglected to get some water from the house.

Aileen was feeling a little woozy and decided to go to the house and bring back some water.

A few minutes later, the Chief called and Gwen answered her mobile. "We found them," said the Chief, "they were a group of high school miscreants who had been listening to the rantings of their parents regarding the LGBTQ and decided to take matters into their own hands.

They will be charged with several offenses and face the possibility of jail time, especially the idiot carrying the gasoline can."

"We want them prosecuted to the full extent of the law. Does that mean we can relax again?" asked Gwen.

"We believe these are the only real threats, but we can't rule out other prejudiced and narrow-minded individuals, so please just be cautious as you have attracted a great many admirers with your choice of lifestyle but also a few critics or those who hold negative opinions."

Gwen hung up, but this news, the heat of the stable and all the things that have happened over the past few months as well as the decisions that need to be made, just overwhelmed her, and she fell to the ground in a faint.

Chapter Eighteen

Aileen returned to find Gwen on the floor of the stable, and she immediately checked her for breathing or injuries. Once she knew she was breathing and not injured, she called the clinic and had Amelia send the ambulance. Amelia said, "Keep her warm and get her head off the cold floor, but try not to move her as there could be internal injuries. We will be there as soon as we can."

The ambulance arrived and they checked Gwen over quickly and then put her on a gurney and transported her to the clinic. They called Dr. Montgomery and explained what was happening and he agreed to meet them there.

Amelia told Aileen she could follow behind and meet them at the clinic and probably contact her parents to make them aware of the situation.

Aileen was in a panic. "*What the hell had happened between the time she went to the house for water and her return?" she was trying to weigh what might have happened.*

At the clinic, Amelia and the doctor checked her over and administered a saline drip to keep her hydrated. They said it looked like she was physically and mentally exhausted and the heat had caused her to faint. The doctor said, "We will give her a complete physical and make sure all is well, but I wouldn't worry too much as she seems to be in good condition but just worn out. We are going to keep her here overnight for precaution and observation and you should go and get her a housecoat, slippers, and a change of underwear."

"I will be right back," Aileen was sobbing. "I will call her parents and let them know what is going on."

She called Louise Provost at her office and explained that Gwen had fainted from the heat and was at the clinic for observation. The Dr. believed she was a victim of the heat and humidity in the stable, but they are going to keep her overnight for observation. She told Louise that she was going back to the house to get some clothes and underwear for Gwen and would meet them back at the hospital, but not to panic.

Louise responded, "We will get there in a few minutes and will wait for you to return."

They were all assembled in the waiting room, hoping the Dr. would be out soon to give them some good news.

Dr. Montgomery came out of the room and spoke with them all saying, "She will be fine as it is just a case of fatigue and stress combined with the heat. She tells me that the two girls have been going through immense stress, strange situations, people killed on the property, and Gwen attacked at the show. This would be too much for anyone. Please let her rest here tonight, she will be able to leave tomorrow but will still need to be required to rest for a week. Aileen, do you have some clothes for her?"

"Yes, right here. "Aileen replied, "as well as some toiletries."

"Please bring them in here," said the Dr.

"I understand that you and Gwen are married, and although I don't thoroughly agree with it, I respect your right to live your lives but I wanted to bring this to your attention as her life partner and let you handle the informing people."

"What is the problem," Aileen was shaking. "Is she going to be all right? We need each other to survive."

"Did you know that she is about 6-8 weeks pregnant?" Dr. Montgomery said abruptly.

"No, I did not. How can that be? We talked about a family but decided to put it off for a year, and I know she has not been with a man."

"You had better speak with her first and let her explain." Replied the Dr.

Aileen was concerned about confronting Gwen about the news while she was in the hospital resting, but this couldn't wait as it would affect their whole lives. She entered the room quietly and tiptoed over to the bed. Gwen appeared to be asleep but rolled over and reached out for her when she saw her and said anxiously. "I am sorry for causing so much upset over being tired and hot, but I will be all right tomorrow and back at the stable with you."

"No, you won't. You are going to be in **our** house and resting for a week like the doctor ordered." Aileen was trying to hide her anger and disappointment. "When were you planning to tell me about the baby before or after the birth?" Aileen couldn't hide her feelings any longer.

"I need to explain my darling so please just give me a few minutes to gather my thoughts." Gwen was shaking and sobbing. "It was during the time that you and I were fighting and deciding on whether or not we were going to stay together. My parents were still nattering at me about our relationship and then the attack, I felt very alone and lost. I went to Dr. Roxanne and we discussed the procedure; even though she cautioned me, I decided to go ahead and at least have someone of my own. I know it was stupid, but I was lost without you and the world seemed to be closing in on me. Can you ever forgive me my darling?"

"There is no need for me to forgive you my precious, but can you forgive me for causing you such pain and sorrow?" Aileen was crying now as well. "All this was already planned and this only changes our timeline and priorities. We need to regroup at the house and consider telling our parents as this might be a bit of a shock, even more than the wedding. There are lots of things happening and we will just re-align assignments, maybe give Aubrey more responsibility, and even get another assistant. I couldn't be happier."

She sat on the edge of the bed and gave Gwen a long, passionate kiss and embrace.

The next day, Aileen went to the clinic and retrieved Gwen, took her back to their new home, and got her settled in bed with some tea and a book.

"Listen, sweetheart, I have to go to town to get us some groceries as our cupboards are bare, and while I am there, I will pick up some oats and apples as we have been ignoring our charges lately," Aileen stated.

"Go and drive safely as you are going to be a parent remember." Gwen was giggling.

Aileen also said she would call Aubrey and have him come to the house for a business chat and make sure he was on board with all the changes.

Aileen picked up enough foodstuffs to get them through the next two weeks and loaded up the back of the truck with the oats and apples. Just as she was getting ready to leave, her father pulled up alongside her and wanted to chat. She would try to hide the pregnancy from her father, but he could always tell when she was lying, so this would be hard.

"HI Princess, how are things going lately and do you need me to help with anything?" James queried.

"Nope, we are fine and just getting settled in together in the house." Aileen was blushing and she wanted to change the subject. So, she asked. "How is Mom, and how are the modifications coming along?"

"The mods are coming along fine and should be done by the end of the week, but Mom is not doing so well according to the Dr. the disease is progressing faster than expected and Mom knows that things are getting worse. I am very worried and could use your help.

"I will get over it when I can, but everything has been so confusing. Gwen needs more attention than we had planned. I have a man coming to give us a quote on a canvas building to cover the jumping arena, one coming to assess Bluster and Artemis for fertilization, and I need to get another person out there to help out."

"I understand," Her father replied. "But please get over to see your mom in the next day or so. By the way, how is Gwen doing with the pregnancy?"

Aileen was startled and stammered, "How the hell do you know about that, and who else knows?"

"I just happened to be in the hospital after you called to tell me about her fainting and when the Provost family left, I overheard Amelia and the Dr. talking and remarking about how well it was going for six weeks and you don't have to be a rocket scientist to connect all those dots."

"She is doing great and just got overheated. You can't tell anyone, Dad, please, as I haven't called her parents yet and they need to hear this firsthand, OK?"

"Can I tell you, mother?" He pleaded.

"Yes." She answered, but that is all.

"I have to go," he said, "They have found something strange at the Harrelson farm and we need to investigate. You know I bought the place, right?"

"No, I didn't, " she was confused, "why?"

"Going to use the land for grapes and hops and put a small store, taste testing, and office out front. I will lease the land back to the company."

"Wow," she gasped. "Things are changing as you have our land at the stable for barley and malt and your farm for grapes and hops. Phoenix Resources is truly using all local materials."

Gwen was anxiously waiting for Aileen to return with the groceries and any news from town.

Chapter Nineteen

Aileen left the larger items in the truck until tomorrow, and as she was putting the groceries away, she turned to Gwen and said, "We need to call your parents and let them know what is happening as you know how gossip can spread in town and we don't want them to hear about the baby from someone else now do we?"

"No," replied Gwen, "I will call them right away."

Gwen called her parents at their office, which they shared between the companies, and asked to see them right away.

Her father answered the call and suggested she come directly after lunch, around 1 p.m. and they could speak in private.

As Gwen was leaving to go and see her parents, Aileen was meeting with Aubrey to make him aware of the situation and go over the plans for the medium term so she would be staying at the house and Gwen could call if she needed support.

Matthew and Louise were waiting when she arrived and she began the conversation as truthfully but gently as they were probably going to be upset and concerned.

"Mom and Dad, we have some news and I wanted to tell you myself. We are going to have a baby as I am 8 weeks pregnant and doing well."

"What the hell are you two thinking?" Louise was visibly shaking.

"How could this happen?" asked her father.

"It is a long story, but we were planning to wait until next year, but the situation changed and I went to Dr. Roxanne and had the procedure completed." Gwen was trying not to cry.

"This was Aileen's doing, wasn't it?" Her mother was angry.

"No." Gwen was getting defensive. "This was all my initiative, but regardless, we are over the moon happy and planning our future together with the baby."

"You are both kids." Replied Louise.

"What age were you and Dad when you were married and had me?" Gwen was going on the attack. We have a business that is doing well and a place to live, thanks to you. There is support at the stable, and we are thrilled about the opportunity to be parents. If you are not willing or don't want to be supportive grandparents, then so be it. But this is happening, so you need to be on board as we will look to the two of you for guidance. I have said my piece and now I need to get back to help with the light-duty chores. Goodbye for now."

Gwen arrived back at the house and Aileen was waiting for her on the porch with a big smile.

"Well, how did it go sweetheart?" Aileen enquired.

"Not well, "Gwen answered as she burst into tears. "My mother was terrible and had some nasty things to say about the pregnancy and us. I told them we were very happy together and excited about the coming addition, and if they didn't want to be supportive, then that would be their loss and goodbye."

"I am so sorry my angel, but maybe they just need a little more time to let it sink in and be happy for us. Remember, she was the one who was requesting grandchildren."

"How did it go with Aubrey?" asked Gwen.

"Fine," Aileen replied. "He wants to be a part of our organization longer term and is more than willing to take on more responsibilities and even invest in the business later on. So that was good news."

"Well at least we have each other and the blessing from your dad and I hope your mom will be pleased," Gwen was smiling.

Aileen was quick to respond, "Let's have some dinner and then go for a ride. The girls always make you smile, and then we can relax in each other's arms and do some baby planning."

"I'm in." Gwen giggled. "To hell with the world as we have the 2 and 1/3 of us to be a team."

Their ride to the river was just what they needed to clear their heads and get them re-focused. Once they put the horses away and checked on all the others to ensure they had feed and water, they headed back to the house and found themselves standing in front of the spare bedroom and thinking this would be a great nursery. Thoughts were running through their minds. What about a colour scheme? We have a pile of stuff to purchase and some baby books to read.

As they were discussing layout and decorations, they heard a noise downstairs and went to find out what was going on.

As they reached the main floor, they could hear someone crying and when they rounded the corner, there was Gwen's mom and dad sitting on the sofa. They looked like both had been crying.

Louise broke into tears and came across to hug Gwen and Aileen.

"I am so sorry kids. What I said was stupid, thoughtless, and uncalled for. The two of you have tackled a large initiative with your relationship, the stable, and now the baby has set the bar so high that no one can match you. With all the upheaval, attacks, and killings on the property, you have proven yourselves to be a superior couple and we couldn't be prouder. Of course, we will support you in everything you are doing and will be the most attentive grandparents you could hope for. Over the next few weeks, we can go, if you wish, to the major baby stores in Proctor and do some serious shopping."

"Fantastic idea," said both of the girls in unison. "We will look forward to the trip."

Gwen replied. 'We love you both and want you and the Parkers to be a huge part of our family. We are expecting the news if/when it gets out. It will probably set the town gossip into a whirl and will need a buffer."

Aileen interjected, "Thanks Mom and Dad Provost if I can call you that. Gwen was very upset and I wasn't sure how to make things better, so we went for a ride to our private place. I have spoken with my father and he is supportive, but as you know, my mother, tragically, will be of little assistance and we hope she can hold on for another 6 months or so to see the baby. Let's have a celebration and some wine."

"By the way," said Matthew, "when we were looking through the stable for the two of you, we found this envelope attached to the post by the tack room and is labeled 'For My Girls'. I am assuming it is for you both."

Aileen rescued the envelope, looked over at Gwen, and winked. We know who it is from and wonder what he is going to tell us this time. The note read:

There have been some unusual developments lately and I need to see both of you. BTW, the upstairs spare room will make a super nursery and I will have some surprises for you both shortly. Please stay healthy. Will call in the next day or so, but might even visit.

Angel

Chapter Twenty

James called an executive meeting for all the managers to get an update and to look forward to business upgrades and improvements.

All the managers attended and asked them to bring their quarterly statistics with them as well as any outstanding issues, ideas, and concerns their people might have. All of the departments were doing well, aside from the plastics group, so they needed a brainstorming session to generate ideas to assist those people.

James outlined, "We are leasing land from the equestrian group to grow the special barley and malt needed and locally supplied. Just in case the gossip mill let us down, I have purchased the old Harrelson estate for Phoenix, but in my name, and that land can be used for grapes and hops as well as a small tasting area and sales department. We can use the material from demolishing the old house to help build the new areas.

More importantly, if we are not successful in addressing the plastics department issues, we might need to review suggestions from our great teams for better use of the building and our team members.

We are never going to reduce staff, but we might need to re-assign people to new roles and maybe even send them for specialized training as required.

In the interim gentlemen, please gather information from the brewmaster and wine aficionado as to the types of barley, malt, grapes, and hops we need to have planted and ideas on where/how to start growing these materials and specifically for our production.

On a sadder note, you are probably all aware of the situation with my wife and her ALS. The situation is worsening weekly, so I will need to spend more time away from the office and will be turning over some duties to Grayson and you as managers.

I will be in contact daily but might be unavailable for some meetings. I have the utmost faith in all of you and your decision-making capabilities so just pretend I am on holiday. Try to get back to Grayson with the details of the materials we need and whatever process is required to grow them before the end of the month.

He left the meeting to go to the farm and make sure the people demolishing the house and shed had clear instructions and took their time.

When he arrived, he called the crew all together and clarified, "I want the house taken down carefully and the material saved; we are going to use it again to build the office here. Leave the barn intact, but tear down the shed and level the ground all around it. You are probably going to come across an underground storage area and I want it saved for the wine storage but clean up the whole area and level the ground. Will check in every day with you and if issues arise, the boss has my mobile number."

Just as he was entering his car, his mobile rang. It was Amelia from the clinic and she sounded alarmed and requested he come over right away to discuss Amy. Amy stayed at the clinic for a few hours every day while their house was renovated to help her be mobile.

He arrived at the clinic to find Amelia almost in tears and Amy in her usual room, also crying.

"What the hell is happening?" He asked.

Amelia replied, "We have been monitoring Amy during her stay with us, and the Drs. have come to the understanding that ALS has been affecting Amy for much longer than we had originally suspected, and the progression is much more rapidly taking a greater toll on her."

"So, what are you telling me?" James was nearly frantic.

Amelia replied, "James, we are sorry to say this, but her time is probably much less than we had originally estimated. I know you did a great deal of work preparing for her return and comfort, but she is going to require round-the-clock assistance and you might need to consider a palliative care facility for her well-being and safety."

"No. I can't do that to her, we have been together for ages and gone through hard times. That would be tantamount to abandoning her and I will need to speak with Aileen before doing anything. The house is ready for her and we can move her there tomorrow, I will see if Jo-Anne can increase the amount of time she can spend as well as investigate a full-time caregiver."

Amelia stammered and responded, "I believe you feel you are doing the right things for her, but are you thinking about her welfare or your guilt? Amy is the main factor here and with the limited time she has left, you need to make sure she is safe and comfortable as the ALS worsens. She is going to be unable to do very little for herself and might not even know where she is. Please speak with Aileen as soon as you can and make sure Amy is the focus."

James spoke with Aileen regarding everything that was happening and asked her opinion. She supported whatever decision he made and knew this was a huge burden for him to be carrying and it was not going to get any easier.

After many hours of soul-searching and discussions with the ALS people, Amelia at the clinic, his family doctor, and Aileen, James decided to move Amy to their home for as long as he could manage her safety and care.

The ramp at the front of the house was installed and the downstairs living room and washroom had been converted into the area she needed to be mobile and have the safety and privacy she required.

They had even added a cable and pulley system to attach to the wheelchair ramp, which would assist people in getting up and down the ramp safely. James was constantly trying to plan out short trips and visits to get Amy out and about. To this end, he leased a wheelchair van to enable Amy to get to town and check in on her dress shop.

He communicated to the town council that the van could be made available for other people in need when not being used for Amy, but he would need sufficient warning.

Jo-Anne, the attending nurse, had enlisted the aid of another nurse/PSW to split the time between them, ensuring Amy had someone there 24/7 in conjunction with James being there throughout the evening.

Amy realized this was getting to be a huge financial and psychological burden on James and Aileen and she knew she was going downhill faster than anticipated. Her goal was to be around when the baby was born, so she was willing to do whatever was needed to make it to that day, although this was an improbable target.

Gwen was 3 months along, so Amy knew she needed to be as strong as possible for another 6 months and then they would discuss the MAID solution to have a positive and respectful end to this suffering.

Chapter Twenty-one

When the girls arrived back at their house, they found a surprise waiting for them in the driveway as well as a note attached to the door. There was a truck parked by the front of the house and when they approached the driver, all he would say was, "There is a delivery for Aileen and Gwen and it's to be unloaded and put into the nursery for them."

"Let's see what's in it?" requested Aileen.

The driver threw open the rear sliding door and the contents caused the girls to look startled and thrilled. The truck was full of baby furniture and accessories, a pricey stroller, a bassinette, two chests of drawers, a changing table, and a very fancy crib.

"Who sent all this and do you have a sheet to sign?" Gwen's curiosity was driving her crazy

"Don't know," answered the driver, "we were told to deliver, unload, set it up, and leave and to answer no questions and there is nothing to sign."

"Must have been some of the grandparents," Aileen was guessing.

"OK, then let's get going," Gwen giggled.

An hour later, the truck had been unloaded and all the furniture installed in the, soon to be, baby's room. They tried to give the driver and his helper a tip, but the driver refused saying, "The gentleman handled all the details with the store, so have a great day and congratulations."

Aileen and Gwen stood at the door of the room and started to cry. Who was responsible for all this and then together they said, "the note on the door."

They took the note and sat on the porch before opening. It was brief, and they automatically knew who it was from.

Please accept this small token of my appreciation and for the love and admiration I have for the two of you. Looking forward to seeing how it looks and will be in touch later today. I will use my standard knocking protocol.

Grandpa Andrew

"I knew it," said Gwen.

"Me too." Answered Aileen

Both of the girls were crying and just sat in silence in each other's arms.

An hour later, they heard the standard 5 knocks at the front door and ran to let him in.

"Welcome, our angel, please come in," said the girls in unison. "How can we ever thank you for all the gifts?"

"No need," said Andrew, "It is my pleasure."

"I have some news regarding our mystery people. Thanks to the two of you, the RCMP has arrested several smuggling groups attempting to cross the border to Canada with horse trailers carrying illegal aliens and a large haul of illicit drugs and weapons. Unfortunately, two of them escaped, but the police have descriptions and they should be arrested shortly. The RCMP is going to acknowledge you both for your assistance and will be in touch."

"How did they know they were smuggling?" asked Aileen."When they looked at the horses, they realized the animals were not show horses and were very old and unstable, and there was no tack. Their excuse for bringing them to a show jumping exhibition didn't hold water." Andrew was laughing.

"That should bring an end to all our worries, and I might not see you again for a while, hence my visit."

"We are so grateful for your assistance and friendship. Please don't desert us just yet." Gwen was concerned. "We have the baby to consider now and can't be dealing with those maniacs again without you. Our police are not very helpful."

"You should rest assured I won't be that far away and just look after each other and your pending arrival." Andrew was sounding very positive.

"Can you stay a little while and have a drink?" Aileen queried.

"No thanks." he replied, "there is a matter of some urgency I need to attend to, but you still have my number, right?"

He left the house walking down the driveway, with no sign or sound of a car or any other vehicle, no sounds of helicopters or ATVs.

Aileen looked at Gwen, "Where the hell did he go?"

"I am not sure; he went down the driveway and just disappeared." Gwen looked at Aileen and asked, "Are we sure he's not a ghost as that is just spooky?"

Epilogue

Seven months had passed.

Gwen delivered a beautiful and healthy baby girl and they named her Charlotte Louise. The whole Phoenix community sent cards and flowers, even though some nasty graffiti was again painted on their fence, which they ignored but had painted over. The furniture provided by Andrew was incredible and grandparents and friends filled all the drawers and closets with clothing, blankets, and outerwear until there was no more room.

Amy held on, with almost a superhuman will, for the whole 6 months and was the first of the grandparents to hold Charlotte, albeit with some assistance. Her eyes were filled with tears and she was visibly shaking, but she was very happy.

Just 4 short weeks after first holding Charlotte, Amy passed away peacefully in her sleep, and the whole community was in mourning. James had completed all the refurbishments to the house and he and Aileen decided to use the house for other families in need. They named the facility after Amy and had a dedication ceremony with the mayor, council members, townspeople, and ALS people in attendance.

They named the facility – **Amy's house for people in need.**

Artemis and Bluster had been inseminated and produced two Trakehner foals; the girls named them Minerva and Cassiopeia, keeping the tradition of naming them after goddesses. Their stable was now full again and Aubrey was doing great as the new General Manager, allowing Aileen and Gwen to address the training and the riding schools.

Contracts had been signed to revamp the paddock areas, install the massive cover over the practice arena, and allow the equestrian facility to handle some county equestrian events, signaling their entrance into the larger circuits.

Phoenix Resources was thriving with the new winery addition on the old Harrelson property (now owned by James/Phoenix) completed and being put online. The brewery doubled its output and even the plastics division was holding its own by accessing new 5-year contracts from the automotive industry.

The two Phoenix entities joined forces to build a large greenhouse, assisting with the starting of new cuttings for the vines and hops. They built a small shop at the front to sell flowers and plants and soon would open a full nursery with a store downtown.

Andrew visited Gwen at the hospital to say congratulations, kissed her on the forehead, and then disappeared again. Before he left again, he tacked another note to the stable and read, "*Congratulations my two equestriennes on your new addition to your family and business success; still here if you ever need me, you are my heroines. --- Andrew.*"

The Phoenix had risen again to help people in need.

The End